Insect Flight

An Insect-Eye View of Flight Behaviour in the Wild

John H. Brackenbury

PELAGIC PUBLISHING

First published in 2026 by
Pelagic Publishing
20–22 Wenlock Road
London N1 7GU, UK

www.pelagicpublishing.com

Insect Flight: An Insect-Eye View of Flight Behaviour in the Wild

https://doi.org/10.53061/KWST1821

A CIP record for this book is available from the British Library

EU Authorised Representative: Easy Access System Europe – Mustamae tee 50, 10621 Tallinn, Estonia, gpsr.requests@easproject.com

ISBN 978-1-78427-626-3 Pbk
ISBN 978-1-78427-627-0 ePub
ISBN 978-1-78427-628-7 PDF

Cover image: Painted Lady *Vanessa cardui* flight cascade
Page ii image: Spotted Longhorn beetle *Rutpela maculata*

Designed and typeset by BBR Design, UK

5 4 3 2 1

Printed and bound by Short Run Press Ltd, Exeter EX2 7LW

FSC
www.fsc.org
MIX
Paper | Supporting responsible forestry
FSC® C014540

Contents

	Preface	vii
	Acknowledgements	viii
1.	**How Wings Work**	**1**
2.	**Flight Control and Hovering**	**52**
3.	**Insects, Birds and the Origins of Flight**	**60**
4.	**Butterfly Flight**	**66**
5.	**Flight Behaviour**	**107**
6.	**Male Flight Behaviour**	**129**
7.	**Mating Flight**	**147**
	Index	165

To access the accompanying video clips, go to the author's YouTube channel:

https://www.youtube.com/channel/UCt5n0EAMyluXkSNPjGbPVQg

Egyptian Grasshopper *Anacridium aegyptium* in flight.

Preface

The study of insect flight has traditionally been the preserve of those interested in the physical mechanisms of movement, an area of biology included within the field of human and animal biomechanics. Science can now explain with a fair degree of confidence how an insect can fly, but fortunately that has not erased the mystery and wonder in people's eyes when, after perhaps a long, dreary winter, they spot the first Red Admiral *Vanessa atalanta* emerging into the spring sunlight. This is no surprise because, while we are used to walking, running and even occasionally jumping, there is nothing intuitive about flying: things that fly live in a world of their own; they symbolise the freedom of the skies and often, when we see them, our spirit flies with them. What does it feel like flying, one asks?

In one sense this brings us to the subject of this book, which is to view flight not just as a locomotory phenomenon but as an aspect of behaviour. The question becomes: not only 'how' does an insect fly, but in what ways does flight serve the behavioural needs of the insect? To make a very crude analogy, biological anthropologists tell us how the human arm evolved from animals that once walked on all fours, but a psychologist might be more interested in the manner in which we use our arms as behavioural tools. Think of the many subtle ways that arms, hands and fingers can communicate feelings and emotions, quite apart from their more practical everyday functions. The 'index finger' does exactly that: it indicates. Instrumental music and dance are the ultimate embodiment of physical movement that is married to emotional expression.

No-one would ever claim that insect wings are remotely the equivalent of human limbs except in the purely anatomical sense as appendages. But I hope that this book at least raises the question. Everyone is familiar with the idea that some insects 'dance', and in some sense most people will understand that the dance, simple though it may seem, is the male insect's way of communicating desire, in this case for a female. The dance is a simple stereotyped routine, of course lacking the complex choreography of human dancing, but based on the same instinctive relationship between rhythmical movement and mental state.

Dancing in insects, like any other aspect of normal behaviour, occurs in the wild and cannot easily be investigated in the laboratory. Yet recording behaviour in the field poses an instrumental problem because of the sheer speed at which events occur. I give an example in the book of a particular dance in which five male butterflies meet, interact in a quite complicated choreographed way, then separate, probably never to see one another again: all within the space of one second. Time frames of this order are typical of most in-flight behaviours among insects. It is impossible to compare the kinds of time frame in which insects act with our own, but on grounds of size alone the speed of events in the invertebrate world must be orders of magnitude greater. Speed is one thing, but much more interesting is the question of how much behaviour – possibly quite complex behaviour – is squeezed into those brief moments of activity. The tool of choice if one is to capture events must therefore be high-speed photography.

I have endeavoured in the book to bring together such skills as I have acquired during 35 years of observing insects in the wild. It has been an almost subconscious process of absorption, re-educating my own eye to glimpse the world of a flying insect through its eyes, albeit with the necessary help of technology. As a companion to the photographs, I have posted a selection of short high-speed video clips online. These are referenced in the text with the acronym HSV and can be accessed via my YouTube channel.

https://www.youtube.com/channel/UCt5n0EAMyluXkSNPjGbPVQg

www.john-brackenbury-photography.com

Acknowledgements

I would like to thank my colleague Roy Barlow, who has worked with me over many years on this project and without whom this book would not have been possible, and Amy McVeigh of Lake Image Systems Limited, who provided technical support. My wife Zena has been a constant source of encouragement on those days when nature was less than cooperative!

1 How Wings Work

Flight machinery

Many small insects beat their wings at a rate of several hundred or more cycles per second, faster than the speed at which nerves can send signals to muscles. This apparent contradiction eventually led to the discovery that, while some insects such as dragonflies and damselflies do indeed control their wing muscles 'directly' because of real-time nervous action, the majority depend on an 'indirect' mechanism, which is illustrated in Figure 1.1.

The exoskeleton of each thoracic segment comprises four plates: the notal plate, forming the roof; the sternal plate, forming the base; and two pleural plates making up the sides. These plates are made of tough, elastic material chitin. Inside the wing-bearing segments (meso- and metathorax, making up the so-called pterothorax), the flight muscles are arranged in two sets: dorsoventral, spanning the segment from top to bottom; and longitudinal, spanning the segment from front to back. During flight, these muscles are kept in an active state by low-frequency nervous impulses, setting up a resonance between the muscles and the elastic chitinous plates. When the

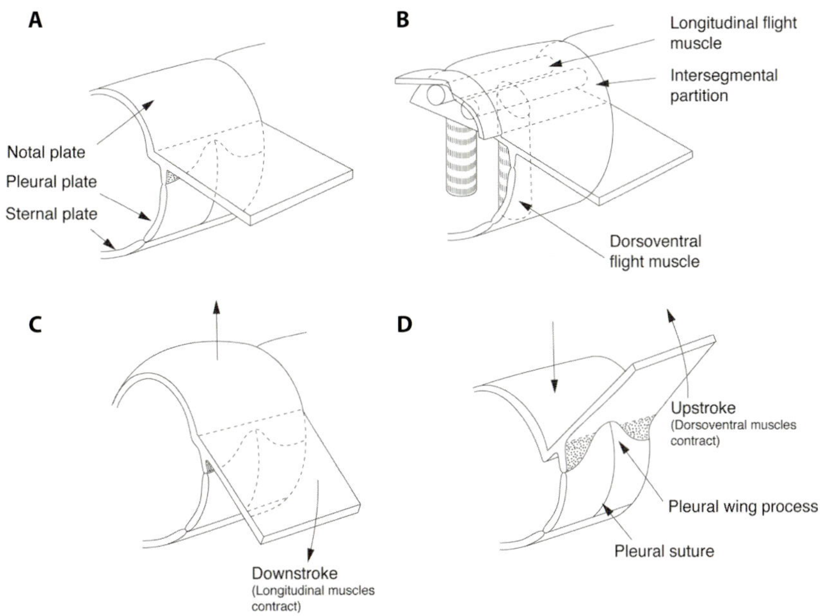

Figure 1.1 The indirect flight mechanism.

Figure 1.2 A recently molted damselfly waits for its integument, or skin, to harden in the sun.

dorsoventral muscles contract, the notal plate is pulled down and the wing is levered up; when the longitudinal muscles contract, the notal plate is pushed up, and the wing is levered down. The scene depicted in HSV.43 showing a dragonfly in the act of egg-laying is considered in detail in a later chapter but incidentally reveals an important aspect of the flight machinery. She is being repeatedly harassed by a male of another species and her only means of defence is to rattle her wings. If you concentrate on the pulsating plate between the bases of the two forewings, you will notice that it moves up as the wings move down, and vice versa. This interesting resemblance to an indirect flight mechanism may be coincidental, since dragonfly wings are driven by muscles attaching directly to the wing bases.

For the drive mechanism of the wings to work effectively, the plates must possess the necessary combination of stiffness to resist buckling and resilience to store the elastic energy. The exoskeleton of a freshly emerged adult is still too soft for the flight mechanism to work efficiently and to achieve full functionality it must undergo a period of hardening, a chemical process like the tanning of leather, in which tight bonds form between chitin and protein molecules. In Figure 1.2 a recently emerged damselfly rests on a water lily, the yellow colour of the flower showing through its soft, still translucent skin.

The attachment of the wings to the thorax is shown in Figures 1.3–1.4, which are drawn directly from high-speed photographs of beetles in flight. The wing base articulates with the body via a chain of very small, hardened

Figure 1.3a Common Soldier Beetle *Cantharis rustica.*

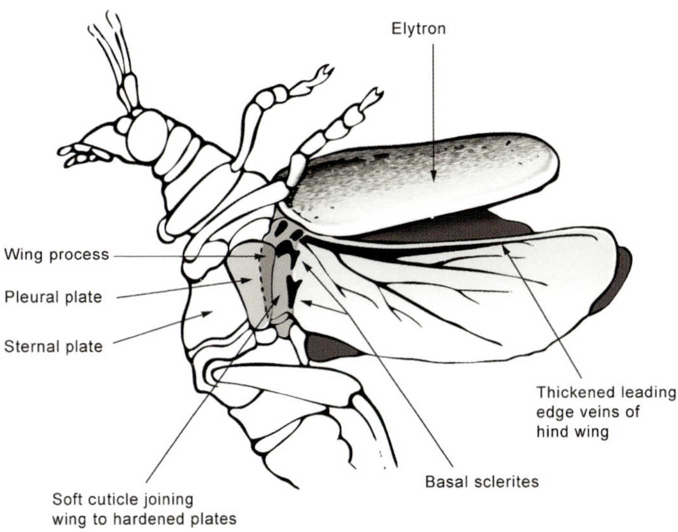

Figure 1.3b Details of the wing and its attachment to the thorax.

Figure 1.4a A longhorn beetle in flight viewed from above.

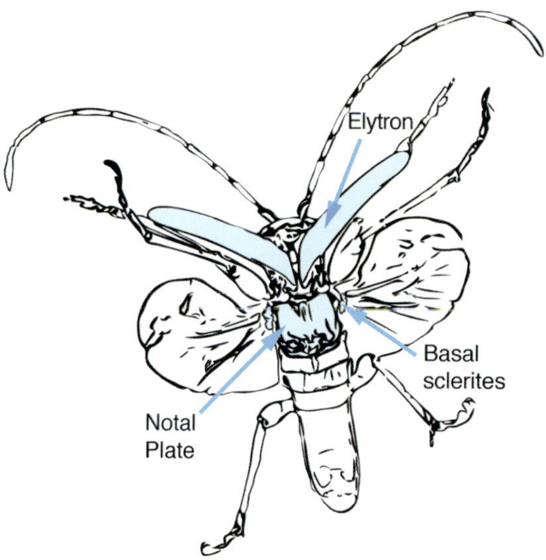

Figure 1.4b Details of the wing and its attachment to the thorax.

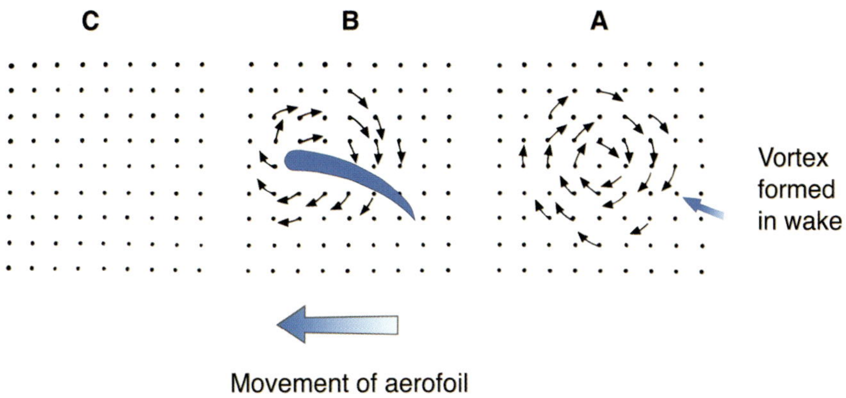

Figure 1.5 Rotational flow around an airfoil.

elements – basal sclerites – embedded in the soft cuticle of the axilla, or armpit, of the wing. The axillary cuticle is impregnated with rubber-like protein resilin, which enables it to absorb the stress imposed by possibly hundreds of thousands of cycles of wing vibration during the lifetime of the insect.

Each wing develops embryologically from a pair of thin shelf-like outgrowths from the notal and pleural plates that eventually become the upper and lower surfaces of the wing.

The space between the outgrowths provides the channel through which nerves, haemolymph, and tracheae enter the wing through a network of veins. The continuous circulation of haemolymph through the veins is essential for maintaining the hydration and elasticity of the fabric of the wing; and some veins become especially thickened to give structural support.

In flight, the angle of attack of the wing determines the amount of lift produced. As an airfoil moves forwards, it creates a circular flow around itself (Figure 1.5) as though it is rotating about its own axis. The greater the angle of attack, the greater the rotational flow and the amount of disturbance shed into the wake. The circular flow is a measure of the lifting force on the airfoil, and insects have evolved special mechanisms for enhancing the rotational flow in ways that lie outside the normal conventions of fixed wing aerodynamics.

To understand the physical principle of how a wing works it is useful to go back more than three centuries to one of the famous laws of dynamics formulated by Sir Isaac Newton: for every action there is an equal and opposite reaction. For the moment, think of a flying insect simply as a body with a mass m moving forwards through the air with a velocity v. The product of these two quantities, $m \times v$, is called momentum, and is the same as Newton's 'reaction'. The law tells us that to maintain its own forward momentum the insect must push an equal amount of momentum into its wake. In an aircraft, this shed momentum is seen in the form of the downwash. To put this into a clearer perspective, think of a rocket propelled by liquid gas. The rocket is moving at a velocity of v and has a mass m. To meet Newton's requirement, it must therefore exhaust to the rear an amount of combusted gas with sufficient weight and velocity to match the rocket's forward momentum. An even closer analogy could be made with a propeller, representing the whirring wings of the insect. In this instance, the propeller accelerates a stream of air into its wake, and the reaction of the propeller blades is the force that drives the aircraft forward.

The force on a wing as it moves through the air can be resolved into two components: lift and drag. In essence these are the same forces that a person would experience while struggling to make headway into a very strong wind: the lift is literally the force that stops them falling as they lean into the wind; the drag is the force they have to work against as they push forward. A thin flat plate moving edgeways through the air will experience neither lift nor drag. If the angle of attack of the plate is steadily increased, both forces will come into play and increase together gradually to a critical point beyond which the drag begins to dominate over the lift until eventually the airfoil stalls. Very loosely, this would be equivalent to a person trying to stand upright in a strong wind, only to be blown backwards.

The maximum lift:drag ratio that can be attained depends on wing shape, or 'aspect ratio'. Long narrow wings are more efficient than shorter, wider wings: hence the design of gliders. The best animal 'glider' is probably the albatross, but high aspect ratios are also found in insects such as dragonflies, damselflies and antlions. At the other extreme lie the broad-winged butterflies, but that does not mean they are less aerodynamically efficient. A butterfly and an albatross are as different from one another in terms of flight technique as it is possible to imagine in any two animals. One glides almost effortlessly on the wind, only occasionally being called upon to flap its wings in earnest; the other is a restless aerobat capable, where necessary, of flipping its body through 90 degrees onto its side and back again with a single beat of its wings. The difference between them lies in the nature of the airflow around the wings. The wings of the gliding albatross work like those of a fixed wing aircraft operating in 'steady' flow conditions. In contrast, the airflow around a flapping wing is not 'steady'; it changes continuously throughout the stroke. This proved to be a stumbling block when scientists tried – and failed – to analyse the aerodynamics of insect flight borrowing from conventional airfoil theory. They had to consider the facts not only that insect wings flap but that (a) the wings on each side of the body can interact and (b) the wings are not rigid, but elastic. These features make the insect wing radically different in structure and function from aircraft wings and form the basis for generating novel kinds of lift from unsteady airflows.

Patterns of wing motion

As wings move up and down their profile is continuously changing due to the interplay between elastic and inertial forces within them and aerodynamic forces acting at their surfaces. The result is that each phase of the stroke has its own characteristics that can be recognised in, for example, high-speed recordings. A wingbeat cycle consists of two 'half strokes': the downstroke, which is usually the more powerful, and the upstroke (Figure 1.6). Between the two half strokes lie the upper and lower stroke transitions, ending at the points where the wing tip reverses direction: stroke reversals or stroke extremes. The angle swept out by the wing tip during each half stroke is the 'stroke angle', which is larger or smaller depending on the effort exerted and is greatest during strenuous flight such as acceleration and take-off. The angle of attack is varied by twisting the wing base: the wing is rotated forwards during the downstroke, in the manner of an arm held out sideways and rotated palm-downwards (pronation). During the upstroke, it is rotated backwards, in the manner of an arm turned palm-upwards (supination). These changes ensure that the wings always meet the air with a positive angle of attack. A hovering insect, for instance, rotates its wings backwards

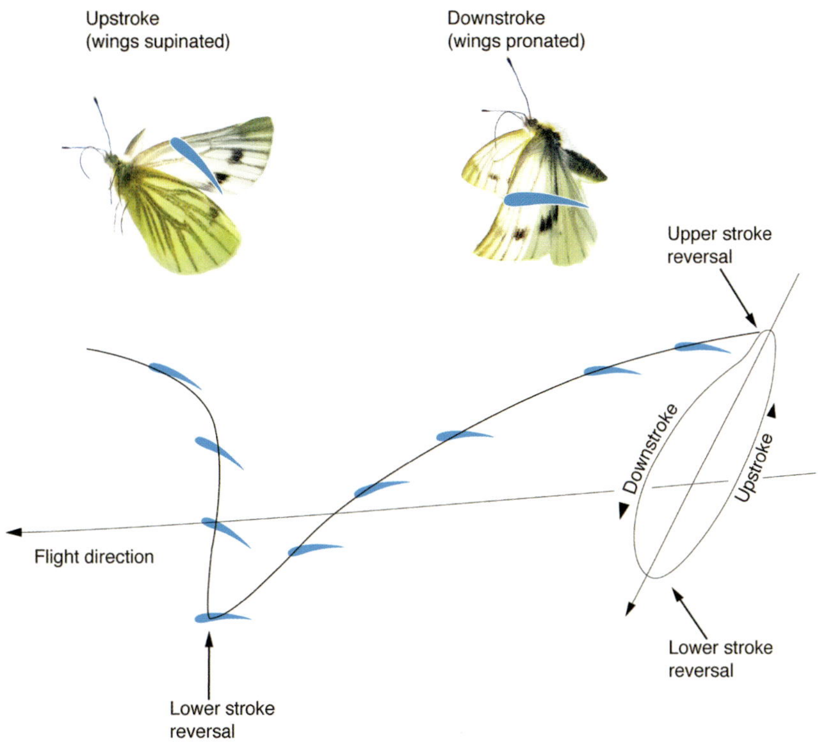

Figure 1.6 Changes in the angle of attack of the wings throughout the stroke cycle.

Figure 1.7 Cranefly. Flight sequence from a high-speed burst of the camera lasting approximately half a second. The insect is flying upwards and backwards from left to right.

and forwards through the same angle during both halves of the stroke. Many readers will have seen this effect in slow-motion films of hovering hummingbirds. At the other extreme, during fast forward flight, the speed of the oncoming air (the relative wind speed) makes it possible to achieve the same amount of lift using a shallower angle of attack, therefore reducing the need for rotation.

From the 'wrist rotation' experiment you will find that your hand can twist through as much as a semi-circle, and people with looser joints can probably achieve more than that. All this rotation takes place not at the elbow but at the wrist, which therefore, in this demonstration, represents the base of the insect wing. In fact, even though the insect wing base is not a mobile joint like an elbow or a shoulder of a mammal or a bird, elasticity makes it just as flexible. The viewpoint from below a flying cranefly in Figure 1.7 shows how the leading edge of the wing rotates through a full half circle between the down (or forward) stroke in image 2 (counting from the right), and the up (or backward) stroke in images 4 and 5.

The wing has two geometrical axes: one along the span, the other at right angles to this, across the chord. Because the fabric of the wing is elastic, it can flex both these axes, either one at a time or both together, depending on the forces experienced at any phase of the stroke. You can easily demonstrate these movements in a length of extendable tape measure. The tape bends more readily towards the cambered side, and insect wings do the same: this is called downward flexure, or spanwise flexure. Transverse flexure produces changes in camber, while simultaneous flexure about both axes results in a propeller-like twist – again; this can easily be seen by twisting the measuring tape. All these effects can be seen to various degrees in different stages of the stroke, but they are most marked at the stroke transitions, when the wing reverses direction and internal stresses reach a peak. The figures that follow illustrate the characteristic features of each half stroke, beginning with the downstroke (Figures 1.8–1.19).

The sequence begins just as the wings are approaching the upper-stroke extreme (Figures 1.8–1.9), at which point typically the right and left wings lie momentarily in contact or near-contact along their upper surfaces (Figures 1.10–1.11). They then peel steadily apart from the leading to the trailing edge in a mechanism that became known as 'clap and peel'. It was discovered by a former professor of zoology at the University of Cambridge, Torkel Weis-Fogh, who found that standard aerodynamic equations failed to account for the lifting force necessary to keep an insect's weight in the air. In short, a bee was too heavy for its own wings. He and his colleagues therefore undertook a search for alternative mechanisms to fill the 'lift deficit'. 'Clap and peel' were one of the first of these to be identified. Subsequently, the

Figure 1.8 Ichneumon wasp. Consecutive stages of wing movement at the top of the upstroke.

Figure 1.9 Bumblebee. Two consecutive stages of wing movement at the top of the upstroke. In (1) the wings are supinated and approaching the upper-stroke extreme. In (2) they are beginning to pronate in preparation for the downstroke.

observational work was given powerful theoretical underpinning by a former student of Weis-Fogh, Charlie Ellington, and the mathematician Sir James Lighthill.

The panels in Figures 1.12–1.13 illustrate in detail the peeling process in the ichneumon wasp *Ophion luteus* and the Cinnabar Moth *Tyria jacobaeae*, and individual stages are shown in the Harlequin Ladybird *Harmonia axyridis*, Red Underwing moth *Catocala nupta* and Large Emerald moth

Figure 1.10 Ichneumon wasp. The wings are in contact along their upper surfaces at the top of the upstroke. In this side view, it is mostly the left wings that are visible.

Geometra papilionaria (Figures 1.14–1.16). As the wings peel apart, a stream of air is drawn over their leading edges, establishing a 'priming flow' for the main phase of lift production that follows once the downstroke gets underway. Clap and peel, or even partial clap and peel, where the wings do not quite make full contact above the body, are especially important in insects with very broad wings, especially butterflies and moths, and in mantids, grasshoppers and crickets, which have large sail-like hindwings. During the downstroke, elastic forces restore the wing to a much flatter profile (Figures 1.17–1.19), and a second type of non-steady lift-generating

Figure 1.11 Heath Potter Wasp *Eumenes coarctus*. The right and left wings have reached the top of the upstroke and are in contact along their upper surfaces. In this side view, only the left wings are visible.

Figure 1.12 Ichneumon fly *Netelia testacea*. Consecutive stages in wing peeling, reading from left to right.

Figure 1.13 Cinnabar Moth *Tyria jacobaeae*. Consecutive stages of wing peeling. (1) mid-upstroke; (2, 3) the front wings are moving slightly ahead of the hindwings as they approach the top of the stroke; (4) the wings have now pronated, are beginning the downstroke and are in the process of peeling; (5) peeling of the hindwings is almost complete; (6) the wings have now completely separated.

Figure 1.14 Harlequin Ladybird *Harmonia axyridis*. Beetles have only one pair of functioning wings, the front wings; the hindwings are modified to form the hard wing covers or shards. Here the front wings have just reached the top of the upstroke and have pronated in preparation for the downstroke. Note that because they are pronated their straight leading edges now point outwards.

Figure 1.15 Red Underwing moth *Catocala nupta*. The peeling process is almost complete, only a small section of the trailing edges of the hindwings remaining in contact.

Figure 1.16 Large Emerald moth *Geometra papilionaria*. The wings are in the final stages of the peeling process, only the trailing edges of the hindwings remaining in contact.

Figure 1.17 Bumblebee *Bombus* sp. The wings are halfway through the downstroke. This is the part of the stroke where the speed of movement of the wings through the air is greatest, their profile is flattest, and they generate lift in a more 'conventional' manner.

Figure 1.18 Caddisfly. Wings almost in the fully down position towards the end of the downstroke.

Figure 1.19 Red Underwing moth *Catocala nupta*. The wings are halfway through the downstroke and have assumed a typically flat profile.

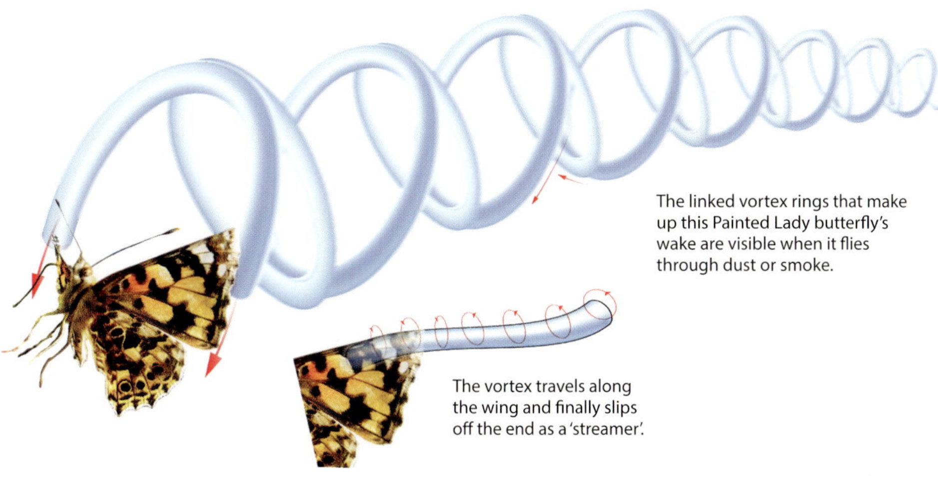

Figure 1.20 The wake of a flying insect. As the wings move through the air a vortex forms along their leading edges, travels along the chord and is shed into the wake as a chain of hoops linked at the upper- and lower-stroke extremes.

Figure 1.21 Caddisfly. Flight sequence from a high-speed burst of the camera lasting approximately half a second.

mechanism comes into play: the 'leading-edge vortex'. As the leading edge of the wing slices through the air, a vortex forms along its upper surface, moves spanwise along the wing and is finally shed from the tip (Figure 1.20). Figure 1.21 is a scenic montage of a caddisfly in flight, constructed from a high-speed burst in a camera fitted with an extreme wide-angle lens.

The lower stroke transition and upstroke phases are illustrated in Figures 1.22–1.31. Even before the downstroke has finished the wings are beginning to twist backward from the base (supinate) in preparation for the next upstroke. Just as in the corresponding upper transition point, elastic forces within the wing now begin to exert their effect. As the inner part of the wing rotates backwards, the outer part continues moving downwards under its own inertia, and the resultant effect is a pronounced downward flexion of the leading edge, together with a propeller-like twist along the span, the latter being especially well shown in Figures 1.26, 1.28, 1.29 and 1.32. In some

Figure 1.22 Ichneumon wasp. Start of the upstroke. The wings have rotated backwards at their bases (supinated) and are experiencing pronounced downward flexure. The insect is in slow forward flight with its body axis tilted strongly down, allowing the wings to move backwards and forwards in the horizontal plane (pseudo-hovering).

Figure 1.23 Cicada. The wings are displaying pronounced downward flexure at the start of the upstroke.

Figure 1.24 Wasp Nest Beetle *Metoecus paradoxus*. Strong downward flexure of the wings during the upstroke.

Figure 1.25 Wasp Beetle *Clytus arietis*. Strong downward flexure of the wings at the start of the upstroke.

insects, such as capsid bugs of the order Heteroptera, downward flexure of the forewing is facilitated by a line of weakness running across the chord (Figures 1.30a and b).

An interesting question arises with respect to plume moths of the family Pterophoridae in which, as the name implies, the wing surface is divided into segments, each consisting of a main vein with a fringe of hairs on each side. Although it is tempting to think that this highly unusual arrangement might

Figure 1.26 Spider-hunting wasp *Sceliphron destillatorium*. Slow, pseudo-hovering flight. The wings are halfway through the upstroke, are strongly supinated and are moving horizontally backward.

Figure 1.27 Common Wasp *Vespula vulgaris*. This rear view of the wings at the start of the upstroke shows the marked downward flexure and propeller-like twisting of the wings.

Figure 1.28 Red-tailed Bumblebee *Bombus lapidarius*. A view from directly below the insect shows the marked downward flexure and propeller-like twisting of the wings during the upstroke.

Figure 1.29 Lacewing. Pronounced downward flexure and propeller-like twisting of the wings during the upstroke.

Figure 1.30a Capsid bug (Heteroptera). The forewing is divided into a toughened area at the base and a membranous area at the tip. The rear part of the toughened area (the embolium) overlaps the leading edge of the hindwing; the front part is divided into a basal section (the corium) separated by a flexure line from a smaller apical section (the cuneus).

Figure 1.30b Capsid bug. Wings at the start of the upstroke. Bending of the forewing occurs at the flexure line between the corium and the cuneus.

Figure 1.31a Plume moth. Seen from directly below (left) and from the side (right) at the start of the upstroke. Note the downward flexure, combined with propeller-like twisting of the wings.

Figure 1.31b Plume moth. Seen obliquely from behind. Note the trailing edge of the hindwings in the finishing stages of the peel.

Figure 1.32 Harlequin Ladybird *Harmonia axyridis*. Flight sequence from a half-second, high-speed burst of a camera fitted with an extreme wide-angle lens.

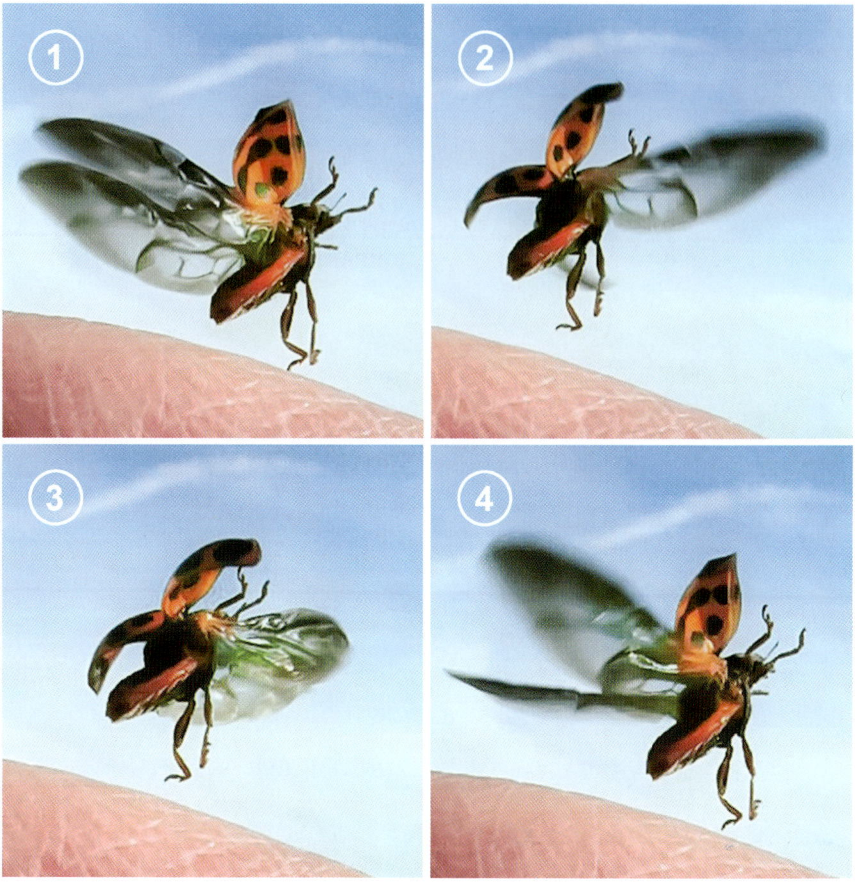

Figure 1.33 Harlequin Ladybird *Harmonia axyridis*. Consecutive stages in a single wingbeat.

be a special adaptation to flight, the dynamics do not support this view. At rest, the segments close together like a fan, giving the wing a thin, stick-like appearance which seems to be an adaptation for camouflage. During flight, however, the segments open fanwise and the wing behaves like a continuous surface. The upstroking wing displays the same downward flexure and propeller-like twisting seen in other insects (Figure 1.31a), and a characteristic peel takes place in the early part of the downstroke (Figure 1.31b).

If, as earlier workers on insect flight discovered, a bumblebee's wings seem paradoxically short for an insect of its size (Figures 1.9, 1.18), the ladybird's seem inordinately long. In fact, they are twice the length of its body, which is why an elaborate folding mechanism evolved for stowing them away beneath the elytra when not in use. The wings execute a stroke angle of 180 degrees and a clap and peel at both stroke extremes (Figures 1.32–1.33). Ladybird flight takes 'unconventional' aerodynamics to its limit, but this may have been a necessary evolutionary development to make up for the loss of aerodynamic function in the hindwings. Figure 1.32 is a scenic montage of a flight sequence recorded through an extreme wide-angle lens.

Warning coloration and preening

In addition to locomotion, insect wings serve several other functions. For example, they are natural solar panels: membranes of large surface area, easily orientated towards the sun, containing a network of vessels that absorb heat and conduct it back to the body. The colours and markings on wings are important in interspecies recognition, courtship, camouflage and defence against predators. The subject of warning coloration has been intensively studied since the Victorian naturalist Alfred Russel Wallace coined the term to describe the use of bright body coloration to advertise to a potential predator that the prey is unpalatable or can defend itself through stinging. It is seen in its most familiar form in the bold yellow and black banding on the bodies of social bees and wasps, potter wasps (Figure 1.11), ichneumon wasps (Figure 1.8), the spider-hunting wasp *Sceliphron destillatorium* (Figure 1.26) and various wasp mimics such as hoverflies, the Wasp Beetle *Clytus arietis* (Figure 1.25) and the Black Yellow Owlfly *Libelloides longicornis* (Figure 1.51). Red and black is also a common colour combination, a conspicuous example

Figure 1.34 Six-spot Burnet moth *Zygaena filipendula*.

being the wings of burnet moths. To all appearances, the Six-spot Burnet moth *Zygaena filipendula* (Figure 1.34) leads a charmed existence. It feeds avidly on the tops of bright, showy flowers evidently with blithe unconcern for its own safety, as it lacks any means of physical defence or method of quick escape. Like other members of the family Zygaenidae, however, it is endowed with potent chemical defences in the form of toxic cyanide compounds. It is commonly assumed that birds are the natural predators of burnet moths and according to the theory of 'aposematic colouring' (the term coined by Wallace), naive birds learn from bitter experience not to attack this insect. The evidence from Figure 1.35 and HSV.1 suggests that the message is not heeded by predatory insects. A Six-spot Burnet moth is suddenly under attack from the air by a Common Wasp worker. The assailant initially tries to dismember the moth by amputating its wings, a tactic regularly employed by wasps to immobilise their prey, but instantly withdraws its mandibles, presumably with a taste of bitter almonds in its mouth, whereupon the moth frees itself quite easily and makes an escape.

Other insects displaying black and red warning colours on the body or wings include the Red-banded Sand Wasp *Ammophila sabulosa* (Figure 2.2), the Cinnabar Moth (Figure 1.13) and the formidable Tarantula Hawk *Hemipepsis mauritanica* of southern Spain (Figure 1.36). Females of this spectacular wasp possess one of the most potent of all insect stings, which

Figure 1.35 Six-spot Burnet moth *Zygaena filipendula*, under attack by a Common Wasp.

Figure 1.36 Tarantula Hawk *Hemipepsis mauritanica*.

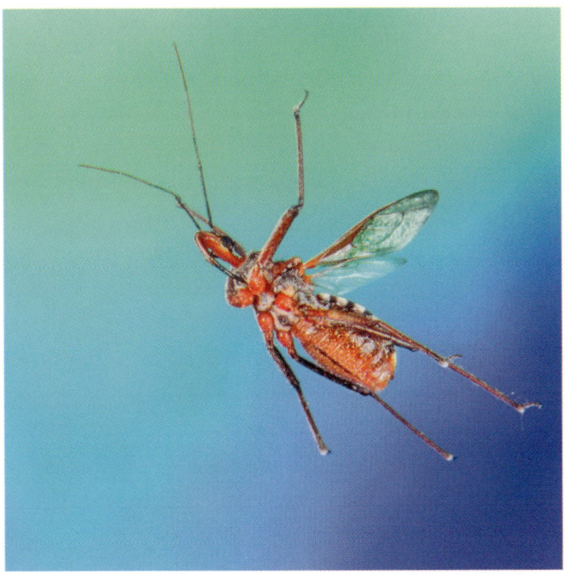

Figure 1.37 Assassin bug *Rhynocoris iracundus*.

is extremely painful to humans, and is used to paralyse the spiders on which the Tarantula Hawk larva feeds. The Tarantula Hawk itself, by contrast, is a nectar feeder.

The assassin bug *Rhynocoris iracundus* (Figure 1.37) is an ambush predator bug in the habit of lurking on the undersides of flowers, waiting to seize unsuspecting bees and flies with its prehensile front legs before impaling them on its fearsome looking 'beak' to inject venom. The bright red and black colouring of its body signals to birds its willingness to retaliate if attacked. It is closely related to the rather menacingly named but much drabber-looking Masked Hunter *Reduvius personatus*, which flies to outhouses and lighted bedroom windows at night in search of small arthropods on which it feeds, including bed bugs. If mishandled, the Masked Hunter can administer a painful bite. Assassin bugs are a very dangerous group, a particularly sinister relative being the 'kissing bug' of Central and South America, whose bite can carry Chagas disease.

The red and black colours displayed on the elytra of various coccinellid beetles, such as the Seven-spot Ladybird *Coccinella septempunctata* and its close relative the Harlequin Ladybird *Harmonia axyridis* (Figures 1.32–1.33) are also associated with chemical defence. Toxins called pyrazines are released from the base of the legs in a process called 'reflex bleeding', and anyone who has handled a ladybird will recall the smear of pungent yellow liquid left on their fingers. The chemical defences of ladybirds exert their effects beyond the domains of birds and insects. When a ladybird becomes trapped in a spider's web the spider recognises the vibrations in the threads, rushes out and prepares to deal with its prey in the normal way by wrapping it in silk and then biting. But in the case of the ladybird the spider applies the silk and then often leaves the inactivated insect to 'wither on the vine'. On coming across one of these victims, even days after it originally met its fate, you will find the telltale orange smear which had repelled the spider; and, if you take a pin and carefully remove the silk wrapping, as likely as not the ladybird will wriggle free and continue on its way.

The Mediterranean Mantis *Iris oratoria* uses eyespots on its wings in a conspicuous, highly stereotyped display known as the 'startle response'. When threatened, it turns to confront the attacker directly, rears back on its hind legs, raises its prehensile front legs in a threatening manner and makes a hissing sound by rubbing its abdomen against its wings: a complex suite of movements made all the more effective by the large quivering eyespots on the hindwings. One imagines the display is even more alarming in flight, where a bird in pursuit of this mantid would be presented with a pair of gigantic, rapidly blinking owl-like eyes (Figure 1.38).

'Flash coloration' is a similar mechanism based on the distracting effect of a sudden display of brightly coloured hindwings when an insect such as a Red Underwing moth (Figure 1.19) or a Blue-winged Grasshopper *Oedipoda caerulescens* (Figure 1.64) suddenly bursts into flight to evade a potential enemy. A bird in pursuit becomes even more confused when the grasshopper lands, closes its wings and, due to the camouflage coloration on the forewings, suddenly disappears. In the Canary Islands, the grasshopper adapts the colour of its body to match the ash-grey and cobalt shades of the surrounding bleak volcanic landscape. On several occasions while walking

Figure 1.38 Mediterranean Mantis *Iris oratoria*.

in this environment I have disturbed individuals, kept a close eye on exactly where they landed and yet could not find them.

When we think of the word 'preening', images of birds and monkeys come to mind, but not insects. Yet regular cleaning is just as necessary for wing maintenance in a fly as it is in a swan. Not unexpectedly, the trait is particularly noticeable in insects with hairy bodies, such as flies and bumblebees. Flies preen with scrupulous care, using the front legs to comb the head and antennae and the hind legs to brush the wings. During their evolution, nesting bees have refined the art of preening into an elaborate system for harvesting pollen from their bodies and storing it in specialised pollen sacs on the legs or the abdomen. It is much less common to catch grasshoppers or beetles in the act of preening and, perhaps not coincidentally, the integument of these insects is often smooth and shiny, and less likely to harbour particles of detritus, while the more susceptible membranous hindwings are concealed at rest beneath the hardened forewings. Preening is not possible in dragonflies and butterflies because their wings are too big and their legs too short. Yet butterflies are particularly sensitive to anything

Figure 1.39 Brimstone *Gonepteryx rhamni* warding off an intrusive fly.

touching the wings. In Figure 1.39, taken from HSV.2, an over-curious fly attempting to land on a Brimstone butterfly is 'cuffed' out of the way with a petulant flick of the wings.

Asynchronous wings and counter-stroking

Apart from the 'true' flies or Diptera (literally, two-winged), all the insects dealt with so far possess two pairs of wings: a front pair and a hind pair, belonging to the mesothorax and metathorax respectively. In most cases the forewings and hindwings are coupled together mechanically, and in effect act as a single aerodynamic surface. The wingbeat is described as 'synchronous', although the forewings always set the rhythm by moving slightly ahead of the hind. Insects in which the forewings and hindwings are not coupled are able to deploy them independently. The two pairs beat at the same frequency, but not necessarily in phase with one another. Even in the same insect, the degree of asynchrony can vary from total, when the wings beat in perfect unison – known as 'parallel-stroking' – to complete or almost complete asynchrony, known as 'counter-stroking'. The most familiar examples of the latter pattern are dragonflies and damselflies. Hovering in the sunlight, their counter-stroking wings give the impression of miniature windmills (HSV.3a and b).

The images in Figure 1.40 are frames of a single wingbeat taken from a recording of a hovering damselfly (HSV.3c). The drawings below illustrate the various movements taking place in the wings at the corresponding stages. By comparing the points at which pronation and supination occur, one sees that the movements of the front and hind pairs are out of phase by almost a whole half cycle. Figures 1.41a, b and c show stages of dragonfly flight just after take-off and, in Figure 1.41c, many of the mechanical features of wings described earlier in the chapter are on display. Both wing pairs have supinated and are moving backwards, although one pair is just beginning the stroke while the other has almost completed it. At this point the wings are positioned in the characteristic 'windmill' configuration, also shown in a damselfly in Figure 1.42. The main part of the wings is almost completely inverted, allowing it to produce lift on what is technically its upper surface. The base of the wing, being firmly attached to the thorax, is unable to invert to the same extent as the outer part: hence the characteristic propeller-like twist. In fact, during these moments each wing not only looks but also behaves like a propeller. The twist in the wing produces a gradation in the angle of attack from steepest at the base to shallowest at the tip. If you bear in mind that (a) the tip moves faster through the air than the base and (b) at any point along the span the lifting force must be proportional to the air speed multiplied by the angle of attack, then it follows that the twist maintains an even lifting force along the whole length of the wing. It was quite amusing to observe two very clear examples of propeller-like twisting

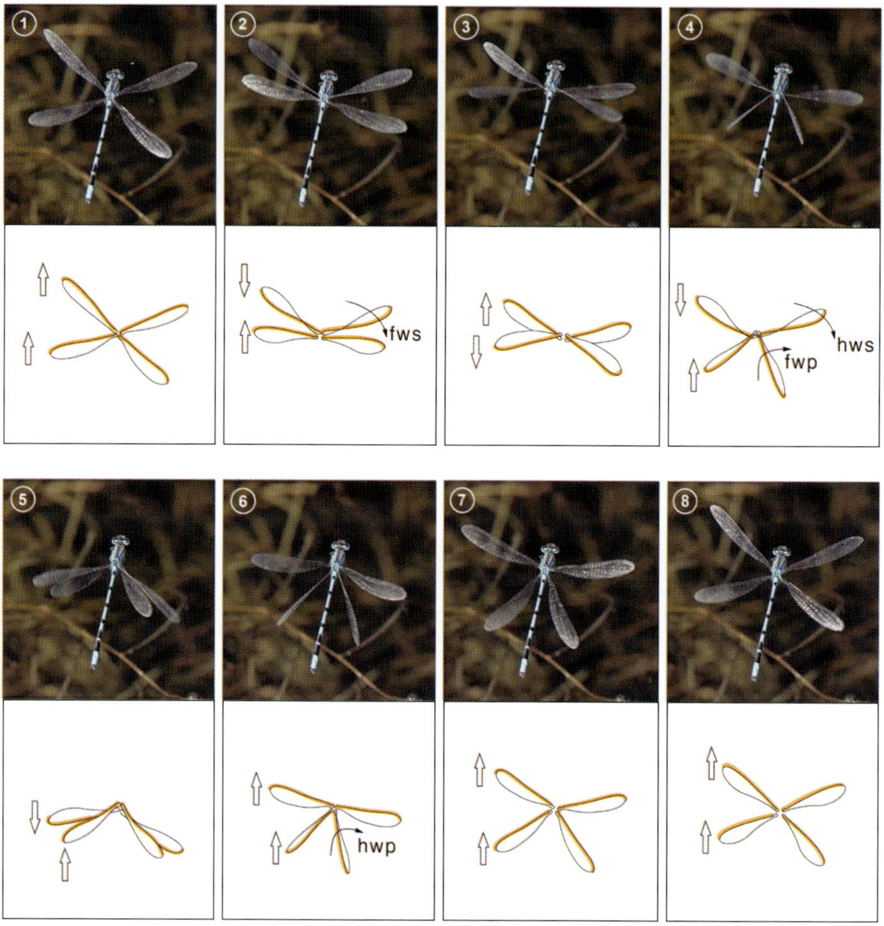

Figure 1.40 Consecutive stages in the wingbeat of a hovering Common Blue Damselfly *Enallagma cyathigerum*.

Diagrammatic interpretation: open arrows, direction of wing movement; wing leading edge highlighted in red.

Key: fws – forewing supination; hws – hindwing supination; fwp – forewing pronation; hwp – hindwing pronation.

Figure 1.41 Common Darter dragonfly *Sympetrum striolatum*. Different types of wing configuration.

Figure 1.42 Damselfly. Wings in the 'windmill' position.

of the wings in the shadows of a dragonfly and damselfly hovering in the sunshine (Figures 1.43–1.44, HSV.3d).

Counter-stroking is suited to hovering and slow flight, but for rapid acceleration and take-off dragonflies and damselflies both adopt parallel-stroking. In between there are many variations, and the insects seem to switch from one rhythm to the other in the space of a single wingbeat. Figure 1.45 outlines a complete wingbeat cycle in a Banded Demoiselle damselfly *Agrion splendens* taken from HSV.3e, and in this case the forewings are beating only slightly ahead of the hindwings. In HSV.4a–d female Brown Hawker dragonflies *Aeshna grandis* employ parallel-stroking for the first few wingbeats during take-off, then transition to counter-stroking.

Why is the technique of counter-stroking particularly suited to hovering and slow flight? The answer lies in the interaction between the forewings and hindwings as they cross in the middle of the stroke. Figure 1.46a is a diagrammatic side view of the hovering insect at stage 5 in the sequence shown in Figure 1.40. The forewing is moving forwards, and the hindwing is moving backwards relative to the body. As the leading edges draw close, their circulatory flows reinforce one another and lead to enhanced lift production.

Figure 1.43 Propeller-like twisting of the wings in the shadow of a hovering dragonfly.

Figure 1.44 Propeller-like twisting of the wings in the shadow of a hovering damselfly.

The thorax of dragonflies and damselflies is tilted back from the head, giving it a noticeably 'humped' appearance when viewed from the side (Figure 1.47a). The upper (dorsal) surface of the thorax faces to the rear, tilting the plane of the wing bases upwards compared with the normal condition. This anatomical reconfiguration of the thorax, given the term 'skew', automatically shifts the stroke plane of the wings closer to the horizontal: these details can be seen from a careful examination of the Willow Emerald Damselfly *Chalcolestes viridis* in HSV.3a. It was originally proposed that skew thorax evolved as a take-off aid in large species of hawkers that typically roost hanging vertically from twigs, but I suspect the main advantage was the resultant increase in hovering efficiency, which follows naturally from the altered stroke plane.

Although dragonflies and damselflies belong to the same order of insects, Odonata, and both are aerial predators, their hunting methods are different.

Figure 1.45 Banded Demoiselle *Agrion splendens*. Consecutive stages in the wingbeat cycle. (1) Upstroke/forward stroke: forewings and hindwings both moving forwards; (2) forewings and hindwings at the upper-stroke extreme; (3) downstroke/backward stroke: forewing supination; (4) downstroke/backward stroke: hindwing supination; (5) end of downstroke; (6) start of upstroke/forward stroke.

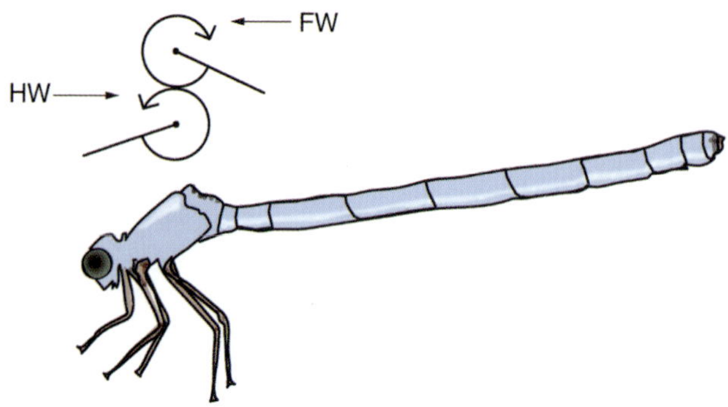

Figure 1.46a Rotational airflow enhancement in the wings of a damselfly during hovering flight. In this diagrammatic side view of the insect, the forewings (FW) and hindwings (HW) are shown in section, their leading edges indicated by a dot. The forewing is moving forwards; the hindwing is moving backwards. Circular arrows indicate the rotational flow around the leading edges.

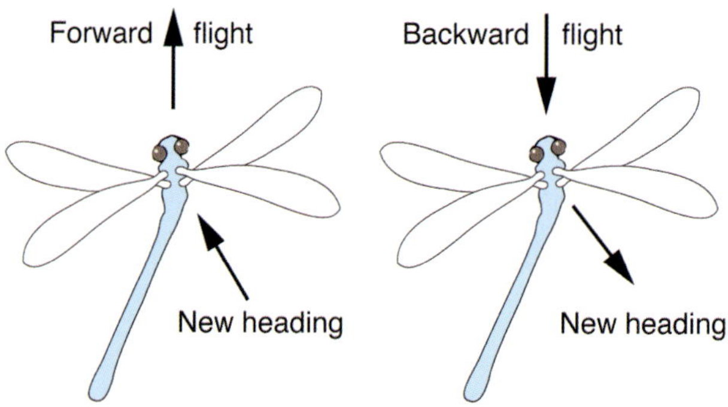

Figure 1.46b Turning manoeuvre in a dragonfly. The insect is viewed from above and has bent its body to the left. If it is moving forwards, the bending of the body will produce a turn to the left; if it is moving backwards, the bending will produce a turn to the right.

Dragonflies are faster and more agile, and capture prey on the wing, whereas damselflies tend to hover around vegetation plucking small insects from leaves, although more robust species such as the Red-eyed Damselfly *Erythromma najas* can capture prey as big as themselves (Figure 1.47b). The Banded Agrion normally has an unhurried, fluttery flight style, but on one occasion I was surprised to encounter an individual plunging into the midst of a column of dancing mayflies, at times flying vertically up and down in pursuit of its prey. Figure 1.48 shows a species of darter dragonfly at the point of take-off, while the scenic shots of a darter and

Figure 1.47a, b, c Damselflies with prey. From (a) the thorax slopes backwards and its upper surface, to which the wing bases attach, faces to the rear of the insect.

Figure 1.48 Darter dragonfly *Sympetrum* sp. at the point of take-off. In this view from directly below, the hindwing is nearest to the camera, and the forewing is viewed through it.

a Four-spotted Chaser *Libellula quadrimaculata* in Figures 1.49 and 1.51 were taken using a half-second high-speed burst of a camera fitted with an extreme wide-angle lens.

I am familiar with the dazzling aerial displays of the oddly named Black Yellow Owlfly (Figure 1.51) as it pursues its prey high above the meadows of the Picos de Europa mountains in Spain, its yellow and black wings shimmering in the sunlight. I always think 'Tiger Fly' would be a better name for them. Video clips HSV.6–8 give an idea of the range of wingbeat rhythms that the insect has at its disposal, and you will notice how the forewings always lead the hindwings in their motion. Sometimes the broad, paddle-shaped hindwings seem barely to move, but because of their size they are arguably still producing lift, albeit passively. The unmistakable Grand Antlion *Palpares libelluloides*, a close ally of the owlfly, also hailing from the warmer parts of Europe, is shown in flight in Figure 1.53, its wings captured in the windmill configuration. And the Common Scorpionfly *Panorpa communis* demonstrates ventral flexure and propeller-like twisting of the

Figure 1.49 Darter dragonfly *Sympetrum* sp. Reconstructed flight sequence.

Figure 1.50 Four-spotted Chaser dragonfly *Libellula quadrimaculata*. Reconstructed flight sequence.

Figure 1.51 Black Yellow Owlfly *Libelloides longicornis*. Consecutive stages in the wingbeat cycle. (1) downstroke; (2) start of upstroke, forewing supination; (3) hindwing supination; (4) pronation and downstroke.

Figure 1.52 Grand Antlion *Palpares libelluloides*. Reconstructed flight sequence. The second image from the top shows the wings in the windmill configuration.

Figure 1.53 Common Scorpionfly *Panorpa communis*. The wings are captured at the start of the upstroke and have just supinated. Note their downward flexure.

Figure 1.54 Common Scorpionfly *Panorpa communis*. The wings are strongly supinated, and a propeller-like twist is evident in the left forewing.

Figure 1.55 Common Scorpionfly *Panorpa communis*. Downward bending of the forewing during supination is located primarily along an indistinct line of flexible membrane.

Figure 1.56 Snakefly *Raphidia notata*. In this reconstructed flight sequence, note the image on the left with the wings at the start of the upstroke. The forewing leads the hindwing, and both are beginning to supinate. The lighting of the image shows how downward flexure of the forewing is locally concentrated along a flexible line of wing membrane approximately halfway along the chord.

Figure 1.57 Pearly Green Lacewing *Chrysopa perla*. Wing dynamics during a single upstroke/downstroke cycle. Note that as in most cases of asynchronous wingbeating, the forewings move in advance of the hindwings. Upstroke: upper row, reading right to left. Right, start of upstroke, forewings and hindwings beginning to supinate. Middle, forewings and hindwings strongly supinated and displaying the 'windmill' configuration. Left, forewings approaching the upper-stroke extreme. Downstroke: lower row, reading right to left. Right, start of downstroke, forewings and hindwings pronated after performing a 'near' clap and peel. Middle, halfway through the downstroke. Left, downstroke/upstroke transition: the forewings have begun to supinate, the hindwings are still pronated.

wings during the upstroke (Figures 1.53–1.54). The flexure is concentrated on an indistinct line across the wing at a point approximately halfway along the chord (Figure 1.55). We see a similar case of localised downward flexure of the forewing in the Snakefly *Raphidia notata* (Figure 1.56). A complete wing cycle in the Pearly Green Lacewing *Chrysopa perla* is illustrated in Figure 1.57 which includes a good example of the 'windmill' configuration (upper row, middle image).

Figure 1.58 Mediterranean Mantis *Iris oratoria*. Wings at the upper-stroke extreme.

Figure 1.59 Mediterranean Mantis *Iris oratoria*. Viewed from behind, showing the hindwings in the terminal stages of the peel.

Figure 1.60 Mantid. Wings captured during the downstroke.

Figure 1.61 Conehead Mantis *Empusa pennata*. In mantids, as in grasshoppers and crickets, the forewings are much smaller than the hindwings. In this example, the forewings appear highly foreshortened since they are viewed end-on at the halfway point of the downstroke. The hindwings lag and are just beginning the downstroke.

Mantids, crickets and grasshoppers also operate the forewings and hindwings independently, but without the strong asynchrony seen in the preceding types. Their large, sail-like hindwings are perfectly suited to peeling (Figures 1.58–1.64), and are often pleated like fans, a smart mechanism for folding them under the forewings when not in use as well as imparting radial strength to the wing during flight, in the manner of a human-made corrugated sheet (Figure 1.64).

Figure 1.62 Locust *Locusta migratoria*. Viewed from directly below. Wings captured in the middle of the downstroke.

Figure 1.63 Large Conehead *Ruspolia nitidula*. The difference in shape between the straplike forewings and the sail-like hindwings is clearly shown. Both are captured at the upper-stroke transition, and the hindwings are just beginning to peel.

Figure 1.64 Blue-winged Grasshopper *Oedipoda caerulescens*. Note the fan-like pleating of the hindwings.

2 Flight Control and Hovering

An animal can fly successfully only if it has safeguards against instability. The principles are the same as those involved in the design of human-made aircraft. The three kinds of instability that need to be controlled are pitch, roll and yaw. An insect without the necessary correction against 'pitch' would tend to nosedive; one poorly corrected against 'roll' would tend to tip onto its side; and one lacking correction against 'yaw' would tend to veer uncontrollably from side to side. Mechanisms against these potential instabilities have been incorporated into the structural design of the flight apparatus over the many millions of years of its evolution and mostly operate without the need for any conscious involvement. An example of automatic adjustment can be seen in HSV.9 and HSV.10. Both the insects concerned, a hoverfly and a honeybee respectively, happened to collide with obstacles as they were taking off, and the impact spun them upside down – a dramatic example of accidental roll. In a matter of a few thousandths of a second the body rolled back to normal, accompanied by a curious series of wobbles. The reaction was completely automatic, the result of an inbuilt anti-roll mechanism.

The fundamental principle of stability in the air is that the centre of gravity of the body should lie below the centre of lift of the wings. When an insect is in the air, think of it as being slung beneath its wings rather like a paratrooper beneath their parachute, or the cradle of a hot air balloon beneath the balloon itself. Any tendency of the paratrooper's body or the cradle of the balloon to sway from side to side in a wind, or even to capsize, is automatically countered. Pendular-like wobbles such as we see in the insects mentioned above are typical in any oscillatory system following a displacement from rest. The bodies of flying insects are built to withstand regular buffeting, and I suspect that minor collisions during take-off are part and parcel of their daily lives, but they happen so quickly that we human beings just do not notice them.

For the purposes of analysis, comparisons are often made between flying animals and aircraft, but in some respects a helicopter provides a better model. 'Hovering' is a case in point. A helicopter uses almost twice as much fuel per unit time when hovering than when flying forward at optimal speed. This is because the rotors of a stationary helicopter must generate all their lift from still air, whereas during forward flight the oncoming wind gives them a head start. The same reasoning applies to insects and explains why sustained stationary hovering is so uncommon: it is costly in energy terms. On the other hand, the flight of many species of slow-flying insects is effectively a

Figure 2.1 Common Wasp *Vespula vulgaris*. Reconstructed flight sequence from a half-second burst of images. Male wasps appear at the end of the season in preparation for the nuptial flight. They are more sluggish in their behaviour and lack a sting.

Figure 2.2 Red-banded Sand Wasp *Ammophila sabulosa*. The upward tilt in the thorax brings the stroke plane of the wings much closer to the horizontal, which benefits hovering. In this instance, the wings are captured at the upper-stroke extreme.

form of quasi-hovering, two examples being the cranefly referred to earlier (see Figure 1.7) and the rather sluggish male Common Wasp shown in Figure 2.1. Genuine stationary hovering of the energetically 'expensive' kind is seen in hawk-moths, hunting wasps such as the Red-banded Sand Wasp (Figure 2.2) and true flies (Diptera), including Muscidae (Housefly *Musca domestica* and relatives) and hoverflies. The dronefly *Eristalis tenax*, familiar to anyone who spends time in the garden, can hold perfect station for many seconds (Figure 2.3, HSV.11a, b) and by no coincidence always seems to hover within easy reach of flowers, from which to refuel as needed. Two equally accomplished hoverers are the Hairy-footed Flower Bee *Anthophora plumipes* and the Dark-edged Bee-fly *Bombylius major*, shown in action in HSV.12 and HSV.13 respectively.

Hovering on the spot is a feat of vision as well as aerodynamics because it relies on precise feedback between the eyes, the brain and the flight muscles.

Figure 2.3 Dronefly *Eristalis tenax*. Hovering. In (1) the wings are at the lower-stroke extreme and are beginning to supinate; (2) wings halfway through the upstroke; (3) wings pronated at start of the downstroke.

In effect, the eyes are the 'spirit – levels' by which the insect maintains a perfectly horizontal orientation. Stationary hovering is just one aspect of the exceptional flight repertoire of the Hairy-footed Flower Bee and the dronefly already mentioned. Towards the end of their respective clips you will see both insects make an exit from the scene using an exquisite 90-degree pirouette on the spot – an extraordinary manoeuvre, different in kind from the usual banked turns (described below), and which still remains a puzzle to be solved.

Bumblebees may have comparatively small wings, but they compensate for this by beating them very rapidly and are among the few insects that can make headway into a stiff wind. Despite their body size, they are proficient hoverers, as we see in HSV.14, featuring a queen Buff-tailed Bumblebee *Bombus terrestris*, freshly emerged from winter hibernation, foraging among springtime snowdrops. The curious onlooker close by, showing impressive hovering skills of its own, is a honeybee, also seen performing in HSV.15. Another Buff-tailed Bumblebee, in HSV.16, hovers briefly as it takes off from Oxeye Daisies *Leucanthemum vulgare*, and totally by chance I happened to capture one squirting out a stream of liquid from its anus as it hovered (Figure 2.4, HSV.17). The high-speed camera shows in detail how surface tension breaks up the tiny stream into a chain of droplets. I doubt if the event lasted for more than just a few hundredths of a second – far too quick for me to have noticed it at the time. Despite initial appearances, I suspect that the bee was excreting not waste but surplus water: foraging bees are known to concentrate the nectar in their honey stomachs before returning to the nest.

Flight performance, in terms of power, manoeuvrability and stamina, varies enormously between groups of insects with widely different lifestyles. The spectacular displays of speed and agility seen in hunting insects will be considered below, and in the next chapter we discover unexpected feats of aeronautics in butterflies. These, however, are the exceptions. If they could be described as the 'aerial display planes' of the skies, far more qualify as humble 'mail planes', performing a workmanlike task with a minimum of fuss. During the summer a vast army of small flies, beetles, bugs, aphids and wasps take to the air and find themselves borne up on thermals and winds, often to great heights, along with a plethora of ballooning spiders, finally to be deposited by chance perhaps hundreds of miles away. Others will remain wafting above pylons, trees and high-rise buildings, their presence indicated by flocks of swifts, swallows and martins trawling through the air in pursuit of these prey.

A late afternoon or evening flight schedule is the norm among many species of weaker-flying insects. Walking along a woodland edge on a warm, still summer evening you might see wasp beetles, click beetles, ladybirds, soldier beetles, cardinal beetles and even black ground beetles (carabids) teetering on the tips of grass stems, tentatively waving their antennae in the

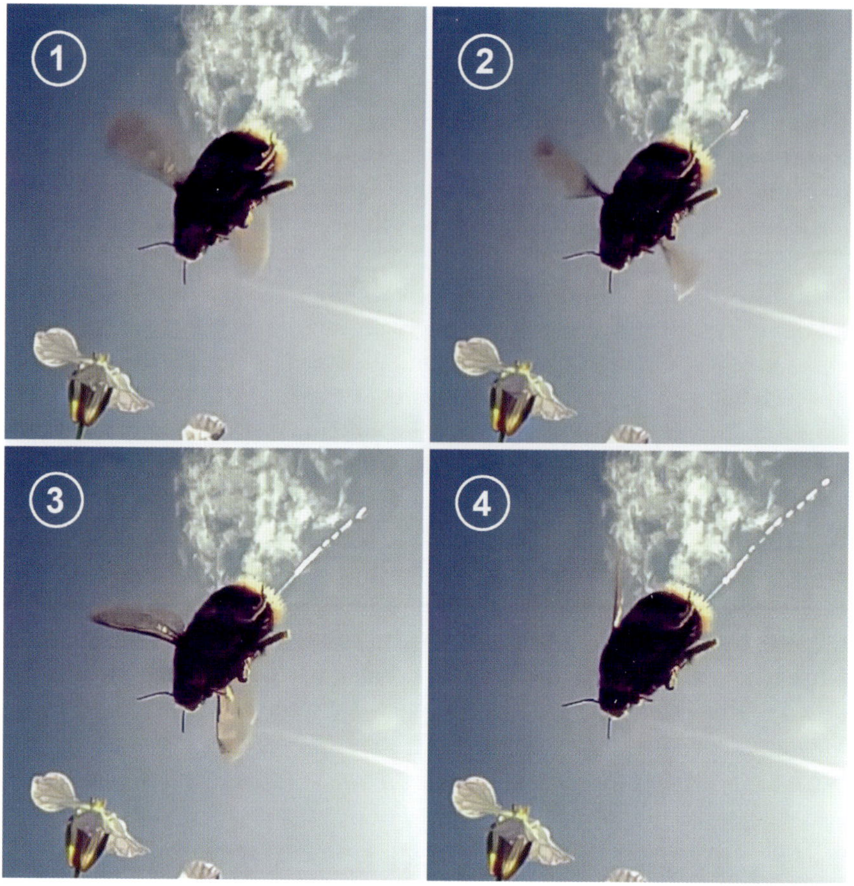

Figure 2.4 Bumblebee, hovering and ejecting a stream of surplus fluid during nectar collection.

air. Shifting their feet and always on the point of launch, they nevertheless continue to hesitate, and you wonder why. One can only guess at the nature of the psychology driving the behaviour of animals so extremely remote from us on the evolutionary scale. In reality, the lingering beetles might be waiting for a 'green light' that we are unable to see: physical signals such as minute traces of pheromones released from females; or faint odours – highly alluring to the insect – emanating from carrion, dung or rotting plant material; or even an almost imperceptible change in the wind direction.

By contrast, there is a much greater intensity and urgency in the behaviour of species that hunt on the wing because action translates directly into meals. Robber flies (Asilidae), dagger flies (Empidae) and dragonflies are some of the most familiar examples. Social wasps, especially hornets, are also voracious carnivores, although they tend to seize insects from vegetation rather than capture them in the air – an example being given earlier in

Figure 1.35. The hunter's efforts are focused on specific targets; therefore, speed, instinct, reflex and agility often come together in explosive bursts of activity. Prowess in the air is one thing, but the hunter must also be able to discriminate between targets that are relevant to its needs and others that would be a waste of energy. This calls for superior capabilities in every aspect of neural activity, from prey detection to analysis of context and tactical execution of the attack.

A very simple experiment sheds light on the first of these capabilities, prey recognition. The typical hunting technique of a darter dragonfly could not be more straightforward: it perches on a prominent piece of vegetation serving as a look-out post, from which it watches for passing prey. When an insect flies by within striking distance, the dragonfly darts out, seizes it and returns to the post to consume it. How does the dragonfly decide on what is suitable prey? Is it based on size, shape, speed of movement through the air or distance? On occasions I have tried to fool the insect by dropping tiny pebbles through the air, but it rarely works. Usually, the insect glances in the direction of the pebble, but without any serious intention of flying out to investigate it. After the first few experiments, one almost gets the impression that the dragonfly finds the whole business a tiresome intrusion into its regular routine.

The gymnastics of a dragonfly as it twists and turns in the air in pursuit of its prey seems at odds with the shape of its body. Of all flying insects, dragonflies and damselflies, with their long wings and 'fuselages', most resemble stately Lancaster bombers. Clearly, such a comparison would be misleading, but it does raise questions of manoeuvrability, such as the method used to change flight direction. Any pilot of a plane would be wryly amused if you suggested that operating a joystick was like turning the steering wheel of a car. Traction in a road vehicle is based on friction between the tyres and the road surface. When a vehicle turns a corner, friction again comes to the aid by stopping it from slipping sideways. In contrast, when an aircraft turns there is no friction to stop it sideslipping, and this is just one of the many variables that a pilot must deal with when carrying out even an apparently simple turn.

The standard technique for changing heading in an aircraft is a manoeuvre called a 'banked' turn. In effect, the pilot uses the ailerons at the trailing edge of the wings to throw the craft temporarily into a condition of controlled instability. Lowering the ailerons on, say, the right-hand-side wing while simultaneously raising them on the left creates more lift on the right wing, making the plane roll to its left. As the plane rolls into the turn, yaw (veering to left or right away from the flight path) and pitch (dipping up or down away from the flight path) also must be controlled, which is achieved through judicious use of the rudder and tail wing elevators. Banked turns are used by insects too, but since insects do not possess tail wings or elevators everything involved in the manoeuvre must be controlled via the wings. The pilot of a plane achieves a tighter turn using a steeper bank angle, but this puts stress on the aircraft frame, and, in the interests of safety, the maximum bank angle is usually restricted to 45 degrees. In the insect flight world, banked turns of 90 degrees or more are probably common and a dramatic demonstration of this can be seen in HSV.18a, where during take-off a female Banded Agrion damselfly rolls through 90 degrees onto her left side (the 'roll in') then back again (the 'roll out') in the space of two and a half wingbeats. Similar versions are performed by a Brown Hawker Dragonfly *Aeshna grandis* in HSV.4a and a Willow Emerald Damselfly *Chalcolestes viridis* in HSV.18b. And in the next chapter we see an even more impressive instance of butterflies achieving the same effect within a single wingbeat.

Finally, there are things an aircraft can do that are beyond the capabilities of an insect, such as a 'barrel roll', in which the plane follows a helical path but maintains the same heading. Granted, such a manoeuvre would be of little relevance to the life of an insect, nor would most of the others that began as tactical manoeuvres in aerial combat. They are exclusively high-speed actions and there are no 'slow' versions. In contrast, the 90-degree 'roll and exit' seen in HSV.4a, HSV.18a and HSV.18b is performed from a stationary position. To make an analogy with human high-jumping, one athlete performs with the aid of a run-up, the other vaults from the spot. The most spectacular example that I have seen of a stationary banked turn was performed by the hoverfly *Myathropa florea* which immediately after take-off momentarily halted, performed a 180-degree roll to its left then exited at right angles to its original heading (HSV.18c). The bluebottle-like Noon Fly *Mesembrina meridiana* executes a less dramatic turn through 90 degrees in HSV.18d. I have already referred to the pirouetting of the insects seen in HSV.12 and HSV.13. In effect, this is a quarter-circle turn on the spot using yaw alone: no banking is involved. It has an odd, clockwork look about it and the mechanism behind it remains a mystery.

HSV.19a shows a dragonfly, captured flying backwards from its perch. The view is from directly below, so by good fortune any sideways bending of the body can be clearly seen. The dragonfly abdomen is very flexible in the vertical plane, as demonstrated in the formation of the copulation wheel and during egg-laying, as the female curls her abdomen forward beneath aquatic vegetation. But sideways bending of the body is less common. Therefore, it was doubly surprising to observe this individual (a) bending its body to its left (to the right from our point of view positioned below) and (b) simultaneously turning to its right. At first sight, this seems paradoxical. As Figure 1.46b (page 40) shows, if a flying insect bends its body to its left, the axis of the thorax becomes realigned to the left, and you would expect the insect to turn to the left. This is precisely what happens in the video HSV.19b of a Common Wasp *Vespula vulgaris*. In normal circumstances the insect would be flying forwards, but from Figure 1.46b we see that if – as in this case – it is flying backwards and bends to its left, it will veer to its right. Such a sophisticated manoeuvre must be under the conscious control of the insect. Adaptive behaviour of this kind is probably not only common but also essential in an insect that survives by hunting in the air, because predation is a two-way process; the hunted may possess evasive skills that match the capture skills of the hunter. Watching dragonflies at work, one gains the impression that capture rates are low and a young dragonfly entering the world with a fully matured skill set for flying may still need to learn lessons on how to hunt. As in various forms of human enterprise, experience will be gained through failure.

Predatory insects are exclusively lone hunters, seeming never to have realised during their evolution the benefits of cooperative endeavour. Hunting in packs increases capture rate and is widely practised among vertebrates such as wild dogs, big cats, wolves and marine mammals but, as far as I know, never in insects or indeed in any other invertebrate. Cooperative activity among insects seems to be confined to specialist social groups such as ants, bees, wasps and termites.

3 Insects, Birds and the Origins of Flight

The origin of flight several hundred million years ago was pivotal in the advancement of insects. Think of the advantages that flowed from it: vastly reduced travel time between destinations; increased dispersal distance through migration; the ability to escape from enemies on the ground and from adverse environmental conditions such as wildfires and flooding; increased foraging range; not to mention the kick-starting of novel patterns of behaviour. The evolution of the plant kingdom experienced a sudden surge as some groups discovered (in a metaphorical sense) that bright, showy flowers could provide the basis for a highly efficient reproductive mechanism. Hence was born the 'floral market', flowers supplying energy-rich food to flying insects in exchange for pollination: one of the oldest and most successful biological partnerships ever to emerge between groups of living organisms.

Despite all the advantages, however, some insects abandoned the air and reverted to a life on the ground. Why, it could be asked, sacrifice all those hard-won benefits? The historical sequence of events is probably lost forever because the fossil record of insect evolution in the very distant geological past is virtually silent. Fortunately, we see encapsulated versions of what it might have looked like in living families of insects that display a range of species all the way from fully winged through partial loss (brachyptery) to virtually complete loss (aptery) of the wings. This trend is shown very clearly in the carrion-feeding ground beetles (Carabidae), where fully winged species rarely seem to fly. Among the wingless species, vestiges of the ancestral wings remain in the form of functionless stubs, as though evolution can never quite dispose of a structure that had given good service for millions of years. Inevitably one is reminded of the human appendix. A direct parallel can also be found in the avian world: the New Zealand kiwi, South American rhea, Australian cassowary and African ostrich all had flying ancestors. Just as in the case of the stubby-winged ground beetles, flightless birds still retain the full paraphernalia of flight, including all the normal skeletal elements and feathers, albeit now only in reduced form.

In discussing this topic, the choice of language is interesting: 'losing' the ability to fly must in practice have been a net 'gain', otherwise it would not have happened. Clearly, there are positive benefits to be gained from secondary loss of flight, and one wonders how many other species are at this moment in the process of 'devolving' back to a terrestrial existence. Evolution in the classical Darwinian sense is like Isaac Newton's 'infinitesimal calculus': it advances by invisibly small degrees. Consequently, there are no easy tests with which to probe this question in real time. But brachypterous or apterous examples exist in virtually every living family of insects, some of the most familiar being glow-worms (Lampyridae, females), oil beetles (Melooidae, both sexes) and velvet ants (Mutillidae, females). Two groups are particularly rich in species displaying wing reduction: geometrid moths and aquatic bugs. The females of the Winter Moth *Operophtera brumata*, Early Moth *Theria primaria*, Mottled Umber *Erannis defoliaria* and Pale Brindled Beauty *Phigalia pilosaria*, for instance, are all wingless. Among aquatic bugs, Water Measurer *Hydrometra stagnorum*, Common Pond Skater *Gerris lacustris* and Water Cricket *Velia caprai* are virtually wingless, while Water Scorpion *Nepa cinerea* and Water Stick Insect *Ranatra linearis*, although fully winged, rarely fly.

Perhaps the most interesting of all are the small aquatic bugs of the genus *Aphelocheirus* (both male and female are usually wingless; Fig. 3.1), possibly the only true 'aquatic' insects in the sense of an organism that spends its entire life – egg, juvenile and adult stages – underwater. Almost all adult

Figure 3.1 The true bug *Aphelocheirus aestivalis* with its small, rounded very reduced wings. (Marianne Müller)

aquatic bugs need to replenish their oxygen reserves by periodic visits to the surface, but *Aphelocheirus* have a evolved a permanent 'gill' through which they exchange gases directly with the water.

The best known of all wingless insects are worker ants (female), and one can clearly see the advantages of winglessness in an insect spending most of its life shuttling back and forth along narrow underground tunnels. On the other hand, male ants do possess wings – unexpectedly large wings, in fact, considering that they use them only once in their lifetime, during the nuptial flight. It is interesting to watch the colony at work when preparing for this unique occasion. Under the careful supervision of the workers, the males are shepherded out from their subterranean chambers into the light for the first time in their lives. They behave like lost sheep, milling around as though waiting for the next set of instructions. If a dark cloud crosses the sun and the air temperature falls, the plan is cut short and the army of males troops back below the surface to await the next burst of warm sunshine.

It is difficult to imagine a more docile insect than the male ant. The easy analogy with sheep is misplaced. Return the sheep to its natural home in the mountains and you see a different animal. In the Picos de Europa mountain range of northern Spain, large flocks of sheep are left to their own devices for days, accompanied only by a very large mountain dog to protect them – but not to guide them. On precarious rocky and grassy mountainsides the flock carves out ladder-like networks of narrow, gently sloping paths linked every so often by short bridging sections. From a distance, the eye sees tiers of white lines moving methodically across the slope: a safe, economic means of passage, devised and executed using raw animal instinct and intelligence.

In terms of biomass, insects probably outweigh all other members of the aerial fauna put together: birds, bats and the prodigious numbers of small spiders ballooning through the air on silken threads. As a plentiful source of food, insects have contributed generously to the evolution of birds and bats, but it is difficult to find any reciprocity. Many large and relatively cumbersome species of flying insects that had enjoyed the freedom of the skies for millions of years would have become easy prey once birds came onto the scene. The giant dragonflies of the Carboniferous era had a wingspan of about 70 cm. The cause of the extinction of these aerial leviathans is still a matter of debate. One of the most popular theories postulates a fall in the level of atmospheric oxygen, the reasoning being that an earlier period of relatively high oxygen concentrations had encouraged the evolution of unusually large insects whose metabolic rates could not be supported by later environmental conditions. Another possibility is predation, but predation by what? Birds seem the obvious candidate, but the geological evidence suggests that the extinction of the giant dragonflies predated the evolution of birds. A third, very imaginative, theory is based on reproductive behaviour, a subject taken up again in Chapter 7.

Figure 3.2 Domestic Pigeon *Columba livia*. Consecutive stages during a single wingbeat.

A comparison of the evolution of flight in insects and birds exposes the fundamental problems facing any animal that exchanges a sedentary life on the ground for one full of new challenges in the air. The competitive pressures driving such a dramatic development are not known, but the odds against long-term success are high, as demonstrated by the rise and fall of the pterosaurs. To use an engineering term, the 'design criteria' for flight are extremely stringent, in regard not only to the anatomy of the flight apparatus itself – the 'airframe' – but also the physiological processes necessarily underpinning it: the systems supplying fuel and oxygen to the working muscles (the 'engine') and coordinating their activity (the 'pilot in the cockpit').

The essential element of the airframe is the wings, and we as fundamentally vertebrate animals instinctively think of wings as modified arms. Following this model, the early pioneers of human-powered flight strapped bird-like wings to their arms and shoulders, leapt off cliffs and attempted to flap their arms like birds. Possibly without being aware of it, they were revisiting the 'ornithopter' depicted in the drawings of Leonardo da Vinci more than 500 years ago. It is touching and inspiring to watch old footage of these brave individuals risking their lives almost literally on a wing and a prayer. To modern eyes their efforts may seem naive, but such men were among the leading innovators of their day and even in their failure their experimental prototypes prepared the way for the fixed-wing machines of today. The fatal flaw in the bird-wing model was not the wing itself – although that would never have worked – but the delivery of power to the wing. The breast muscles of a bird (equivalent to the human pectoral muscles) are massive and are anchored to an equally outsized sternum with a deep, projecting keel. To drive his wings, a hypothetical 'bird man' would require pectoral muscles making up half the entire weight of his body, inserting onto a sternum reaching down to his knees, along with the necessary increases in cardiopulmonary function.

There is no historical record of a prototype aircraft based on an insect model, but the same considerations would apply: regardless of the detailed structure of the wing itself, muscle power will always be the limiting factor. There is a very easy and perhaps surprising way of demonstrating this from the comfort of your own armchair. With your hands on the armrests, try levering your body up. Doing this once is exhausting, and even an athlete with highly developed arm and chest muscles would find it difficult to repeat the exercise more than a few times. But this is exactly what a flying bird, bat or insect must do to maintain its own body weight in the air against the force of gravity.

The primate arm and the bird wing both evolved from the front limb of an ancient reptile, the first truly bird-like reptile being the famous feather-bearing *Archaeopteryx*, regarded as a transitional form between terrestrial dinosaurs (most of which were quite small) and flying dinosaurs. In an anatomical sense, reptiles (which in turn owe their own ancestry to fish) had long been primed to progress from running to flying, when environmental circumstances favoured that transition.

A jointed wing based on the vertebrate plan lends flight possibilities to birds that are not available to insects. An examination of the wingbeat cycle of a bird will show why (Figure 3.2). Note, incidentally, that even when the wing of a bird is fully extended the elbow itself remains partially bent, the triangular gap between the wrist and shoulder joints being filled by a sheet of elasticated skin called the patagium. The sequence of images starts with the wings captured halfway through the upstroke (position 1), which, as in insects, is primarily a recovery stroke. Flexion at the wrist and elbow joints, along with backward rotation at the shoulder joint (supination), 'feathers' the wing, reducing its drag as it moves through the air (position 2). These events are functionally equivalent to supination in the insect wing, and they lead to the same propeller-like twist. The similarity continues as the bird executes a clap and peel at the top of the upstroke (positions 3 and 4) and finally extends the wing during the downstroke (positions 5 and 6). Although the dynamics of the stroke are the same as those of an insect, they are achieved in a different way. Both animals power the wings using muscles originating in the body, but the bird also has muscles and joints inside the wing itself that can be used for fine-tuning wing profile. In contrast, events in the insect wing can be actively influenced only via its pivot at the base. For instance, a swooping falcon makes itself more streamlined by tucking its wings in close to the body. The 'bounding' flight style of finches is based on a similar principle: flurries of rapid wingbeats alternating with periods when the wings are closed, giving the body a smoother profile. Bounding flight is a clever energy-saving device and the nearest equivalent I can think of in insects is the intermittent gliding of dragonflies (HSV.5a and b) and butterflies (HSV.5c). Dragonflies interpose very brief periods of gliding in normal forward flight, whereas butterflies tend to glide, sometimes for several seconds, during descending flight.

4 Butterfly Flight

The flight world of a butterfly

Although comprising only a small fraction of the total number of flying insects in the world, butterflies display so many of the underlying elements of flight behaviour that they repay special consideration. The physical feature setting them apart from other insects is their large, broad wings that beat unusually slowly – depending on species, at about 10 to 25 beats per second. This happens to coincide with the so-called flicker fusion frequency of the human eye – the frequency at which we perceive a sequence of images as a continuous 'movie' rather than a succession of still frames – hence the impression that butterflies 'flutter' through the air, a term we could not realistically apply to any other insect except perhaps dragonflies, damselflies and antlions.

Because butterflies are so physically conspicuous, some aspects of their flight can easily be discerned by the naked eye. For example, in open countryside it is not unusual to see butterflies following tracks, field and woodland edges, hedgerows and other conspicuous ground features that offer a clear line of sight or shelter from the wind – a behavioural trait in keeping with the migratory tendency of the group. The insect's own flight path is rarely straight, but veers quite randomly from side to side, possibly with the intention of confusing an enemy approaching from behind but just as probably as part of normal orienteering behaviour. The 'seeking' flight style allows the eyes to supply the brain with a running commentary on what lies to the left and right of the mean line of flight. This, too, would be consistent with the psychology of a long-distance flyer.

A third and related possibility is distance assessment. It is well known that grasshoppers and crickets exploit the principle of 'motion parallax' by using side-to-side 'peering' movements of the head. For a given sideways motion, a closer object appears to move faster across the background than one further away. I have witnessed this behaviour at first-hand in a large cricket that flew into a tent I had erected on a camping expedition. It ensconced itself on a tripod parked near the wall of the tent and began swaying very deliberately, all the while keeping its eyes fixed on the top of my head. After a few moments it jumped and landed on my brow. I placed it back on the tripod and the same sequence of events followed. This exchange continued for quite a while, until I drew the tent entrance wider and the grasshopper flew back out into the fields.

We use parallax constantly in our daily lives without even thinking about it, especially when we are on the move. As we walk along a path, our eyes register the terrain moving past – the 'optical flow' – as though we were sitting in a train watching the landscape streaming past the window. Trees, hedgerows and telegraph poles at the side of the track shoot by in a continuous blur but hills in the distance hardly move at all. Using this input, the subconscious brain constructs a continuously updated '3D surround' picture of the world. Hence as we walk, relative distances in all directions are computed in real time and we avoid bumping into obstacles. Using the same neural template, sleepwalkers and people suffering from a condition known as 'blindsight' can walk through a room without colliding with the furniture. Blindsighted patients are convinced in their own minds that they cannot see but this is belied by their behaviour.

These examples from human physiology might seem remote from the subject of butterfly orientation, but put yourself into the mind of a bumblebee attempting to land on a flower swaying in the wind or a bird attempting to land on the branches of a tree lashed by a gale. At the same time, consider the care with which a bee lowers its 'undercarriage' in preparation for a landing – all six legs of it: nicely shown by the Hairy-footed Flower Bee in HSV.12. Now think of a helicopter pilot trying to put down their craft on the deck of a ship rolling in a stormy sea. Forgetting obvious differences in scale, the situations are not dissimilar. Both call for maximum coordination between eyes and limbs. The human pilot will have acquired their skills through intensive training, to the point where operating the controls has become instinctive. In contrast, insects are 'born' experts; anything less and they would fall at the first hurdle. It might seem questionable that motor skills of this magnitude could be innate, without any need for learning, considering the exceptional demands that flight places on the nervous system, especially one that contains as few as a million neurons – a drop in the ocean compared with the human brain, with its more than 80 billion neurons. The apparent paradox is resolved when we remember that the computing power of a neural network depends not only on the number of neurons within it but, more importantly, on the number of synaptic connections between them. This biological principle lies at the heart of machine learning and artificial intelligence as well as modern attempts to model human and animal consciousness. The latter science is still in its infancy, understandably driven by interest in the human condition, but it is only a matter of time before attention also turns to invertebrates.

On numerous occasions, I have watched the behaviour of butterflies as they prepare for their overnight roost. Prompted by an internal clock, they usually begin the search an hour or two before sunset, but on a warm, bright evening they may carry on feeding for an extra hour or so. When the light fades and the search finally gets underway, there is an extra sense of urgency in the butterfly's behaviour. For example, a Marbled White *Melanargia galathea* typically patrols restlessly up and down the very same 50 m stretch of woodland edge, hovering here and there to inspect dozens of potential sites and becoming increasingly agitated in its movements. Finally, it opts for a particular grass stem that, to the outsider, looks no different from any other grass stem and, satisfied that conditions are right, it swivels its body through 180 degrees and settles head-down for the night.

In the first chapter we touched on the prey recognition abilities of dragonflies. The pebble-dropping experiment demonstrated in a very simple way that the dragonfly has discriminatory powers. The more you think about it, the more you realise that this must be the case in all multicellular animals, otherwise their lives would be a continuous series of mistakes. Once again, the question arises: how much of this behaviour is innate and how much is

Figure 4.1 Silver-washed Fritillary *Argynnis paphia*. Start of the upstroke; the butterfly is viewed from below as the leading edge of the wings moves ahead of trailing edge resulting in the formation of a trumpet-shaped space between the wings. Simultaneously supination produces strong reverse camber on the wings.

Figure 4.2 Red Admiral *Vanessa atalanta*. Start of the upstroke; the butterfly is viewed from below as the leading edge of the forewings moves ahead of the trailing edge resulting in the formation of a trumpet-shaped space between the wings. Simultaneously supination produces strong reverse camber on the wings.

acquired through learning? Either way, it argues for a certain level of intelligence, in the very broad sense of that word. Curiously, butterflies seem to lie rather further down the scale than dragonflies. The male Silver-washed Fritillary *Argynnis paphia* is a dashing-looking insect with its smart orange wings streaked with brown markings (Figure 4.1), and, as with most male butterflies, its main purpose in life is to chase females. But he is prone to make the wrong choices. Experiments have shown that he is just as likely to pursue a butterfly-sized piece of orange material trailed through the air as he would the genuine article. It is not hard to find similar examples during a walk through the countryside. All males of the British species of white butterflies – that is, the Small, Large and Green-veined Whites and the

Figure 4.3 Comma *Polygonia c-album*. Start of the upstroke; the butterfly is viewed from above as the leading edge of the wing moves ahead of the trailing edge resulting in the formation of a trumpet-shaped space between the wings. Simultaneously supination produces reverse camber and propeller-like twisting of the wings.

Figure 4.4 Dark Green Fritillary *Argynnis aglaja*. Start of the upstroke; the butterfly is viewed from above as the leading edge of the wings moves ahead of the trailing edge resulting in the formation of a trumpet-shaped space between the wings. Simultaneously supination produces strong reverse camber.

Orange-tip – seem to be obsessed with the colour white. Consequently, as a male pursues his quest for a female, he is compelled by instinct to investigate anything in his path that is white and roughly the size and shape of a butterfly. It is amusing to watch his reaction to a handful of white petals strewn on the ground in his path!

Patterns of wing movement

Butterflies make excellent subjects for the high-speed camera, and I include in the adjoining pages examples that illustrate all the salient points of the wingbeat but also showcase the beauty of these colourful insects when they are in their natural element, the air. The photographs are arranged in four sets corresponding to the consecutive stages – lower stroke transition, upstroke, upper-stroke transition and downstroke – although inevitably in a continuously moving system there will be overlap. The most distinctive feature to note about the dynamics of butterfly flight is the part played by the anal lobes of the hindwings, which are so big that they almost completely cover the ventral surface of the abdomen. This is shown in many of the photographs, especially those captured from below the flying insect, such as Figures 4.6, 4.9, 4.13 and 4.21.

The stroke angle is very large – 180 degrees in the case of take-off and climbing flight, where the left and right wings come into contact, or very near contact, at both upper- and lower-stroke extremes. The latter stage is especially interesting in butterflies and for convenience it is taken as the starting point in the description. As the left and right wings begin to separate in preparation for the upstroke, they enclose a flared trumpet-shaped space, the 'mouth' of the trumpet being the leading edge of the forewings, the 'stem' of the trumpet tapering back to the trailing edge of the hindwings (Figures 4.1–4.4). Viewed from above or below at this stage, the butterfly assumes a characteristic kite-like shape (Figures 4.5–4.8). The wing dynamics are in fact very similar to those of a downstroke 'peel' discussed in Chapter 1. As in that case, air circulates rapidly around the leading edge of the wings into the growing space (Figure 4.10). The mechanism only works if there is an airtight seal at the base of the trumpet; hence the significance of the exaggerated anal lobes mentioned earlier which, by folding together beneath the abdomen, act as flap valves (Figure 4.9) isolating the underside of the wing from the upper.

Continuing their upward movement, the wings assume the marked downward bending of the leading edge and propeller-like twist along the chord characteristic of all the insects discussed in Chapter 1 (Figures 4.11–4.24). The propeller effect is especially evident in Figures 4.21 and 4.22. Figures 4.25 and 4.26 show flight sequences recorded in a half-second burst of a camera fitted with an extreme wide-angle lens.

Figure 4.5 Red Admiral *Vanessa atalanta*. Characteristic kite-like appearance of a butterfly seen from directly below during the later stages of the upstroke.

Figure 4.6 Scarce Swallowtail *Iphiclides podalirius*. Characteristic kite-shaped appearance of a butterfly seen from directly below during the later stages of the upstroke.

Figure 4.7 Brimstone *Gonepteryx rhamni*. Characteristic kite-like appearance of a butterfly seen from directly below during the later stages of the upstroke.

Figure 4.8 Red Admiral *Vanessa atalanta*. Consecutive stages during the upstroke. The insect is viewed from directly below. (1) early stage: supination results in strong reverse camber and the formation of a trumpet-shaped space between the wings; (2) halfway through the upstroke: wing profile is now much flatter; (3) later stages of the upstroke as the wings approach the upper transition; (4) the wings are now halfway through the downstroke.

Figure 4.9 Brimstone *Gonepteryx rhamni*. The butterfly is viewed from below at a point halfway through the upstroke. Note the downward flexure of the wings and the anal lobes of the hindwings folded together beneath the abdomen.

Figure 4.10 Large White *Pieris brassicae* viewed from below. Arrows indicate airflow around the wings. Note the anal lobes of the hindwings pressed together and forming a seal against airflow from behind.

Figure 4.11 Painted Lady *Vanessa cardui*. Wings halfway through the upstroke. Most of the downward bending of the leading edge of the wings characteristic of the earlier stage has disappeared.

Figure 4.12 Queen of Spain Fritillary *Issoria lathonia*. The wings have reached the halfway point of the upstroke, with some downward flexure still remaining in the forewings.

Figure 4.13 Large White *Pieris brassicae*. The butterfly is viewed from directly below. Wings halfway through the upstroke. Slight downward bending of the wings still in evidence. Note that the abdomen is concealed from view by the large anal lobes of the hindwings.

Figure 4.14 Large Skipper *Ochlodes venatus*. Early part of the upstroke, with strong downward bending of the leading edge of the wings.

Figure 4.15 Silver-washed Fritillary *Argynnis paphia*. Wings just past the halfway point of the upstroke. Downward bending of the leading edge of the wings.

Figure 4.16 Small White *Pieris rapae*. The insect is viewed obliquely from below at the halfway point of the upstroke. Note the downward arc of the leading edge of the wings.

Figure 4.17 Brimstone *Gonepteryx rhamni*. The insect is viewed from behind and below at the halfway point of the upstroke. Note the downward bending and reverse camber of the wings, and the anal lobes of the hindwings pressed together, concealing the abdomen.

Figure 4.18 Marbled White *Melanargia galathea*. Wings halfway through the upstroke as it flies away backwards. Note downward bending of the leading edge of the wings.

Figure 4.19 Green-veined White *Pieris napi*. The insect is viewed obliquely from behind. Wings approaching the halfway point of the upstroke.

Figure 4.20 Marbled White *Melanargia galathea*. Wings approaching the halfway point of the upstroke. Note strong downward bending of the leading edge.

Figure 4.21 Dark Green Fritillary *Argynnis aglaja*. Wings approaching the halfway point of the upstroke. Note the downward bending of the leading edge, together with reverse camber and propeller-like twisting of the wings. Note also the anal lobes of the hindwings pressed against the abdomen concealing it from view.

Figure 4.22 Painted Lady *Vanessa cardui*. In this dynamic view we see the insect from behind. It has taken off from the ground backwards and its body has been thrown almost upside down. The wings are approaching the halfway point of the upstroke. Supination of the wings combined with downward bending of the leading edge has resulted in pronounced propeller-like twisting.

Figure 4.23 Red Admiral *Vanessa atalanta*. The insect is viewed from below with wings approaching the halfway point of the upstroke.

Figure 4.24 Brimstone *Gonepteryx rhamni*. The insect is viewed from directly above with wings nearing the top of the upstroke.

Figure 4.25 Red Admiral *Vanessa atalanta*. Reconstructed flight sequence taking off from Heather.

Figure 4.26 Brimstone *Gonepteryx rhamni*. Reconstructed flight sequence taking off from Lucerne.

Figure 4.27 Large White *Pieris brassicae*. viewed from behind during early downstroke showing peeling wings.

Figure 4.28 Silver-washed Fritillary *Argynnis paphia*. Viewed from behind during early downstroke showing peeling wings.

Figure 4.29 Great Banded Grayling *Brintesia circe*. Viewed from behind during early downstroke showing peeling wings.

Figure 4.30 Dark Green Fritillary *Argynnis aglaja*. Viewed from behind during early downstroke showing peeling wings.

Figure 4.31 Small White *Pieris rapae*. Viewed from behind during early downstroke showing peeling wings.

Figure 4.32 Scarce Swallowtail *Iphiclides podalirius*. Viewed from directly above the body during the first half of the downstroke. The peeling process is almost complete, only the tails of the hindwings remaining in contact.

Figure 4.33 Great Banded Grayling *Brintesia circe*. Viewed from behind during the first part of the downstroke. The peeling process is completed, and the right and left hindwings have separated.

Figure 4.34 Comparison of the airflow pattern (arrows) around the wings of a butterfly during a downstroke peel (on the left) and an upstroke peel (on the right).

Figure 4.35 Meadow Brown *Maniola jurtina*. This view from below the insect shows the wings fully outstretched in the middle of the downstroke.

Figure 4.36 Large White *Pieris brassicae*. Wings lowered in the later stage of the downstroke.

Figure 4.37 Painted Lady *Vanessa cardui*. The insect is viewed from below with the wings fully outstretched at the mid-point of the downstroke.

Figure 4.38 Common Blue *Polyommatus icarus*. Wings almost completely descended in the late stage of the downstroke.

Figure 4.39 Marbled White *Melanargia galathea*. The view is from above the body and the wings are fully outstretched at the mid-point of the downstroke.

Figure 4.40 Red Admiral *Vanessa atalanta*. Later stage of the downstroke.

Figure 4.41 Meadow Brown *Maniola jurtina*. Viewed from directly below with wings at the halfway point of the downstroke.

Figure 4.42 Marbled White *Melanargia galathea*. Viewed from below with wings approaching the halfway point of the downstroke.

Figure 4.43 Painted Lady *Vanessa cardui*. Viewed from directly below with wings fully outstretched at the halfway point of the downstroke.

Figure 4.44 Green-veined White *Pieris napi*. Viewed from directly below with wings in the later stage of the downstroke. Note the anal lobes of the hindwings folded against one another beneath the abdomen.

Figure 4.45 Brimstone *Gonepteryx rhamni*. Viewed from below during the second half of the downstroke. Note the anal lobes of the hindwings folded against one another and concealing the lower surface of the abdomen.

Figure 4.46 Large White *Pieris brassicae*. Viewed from below with wings fully outstretched at the halfway point of the downstroke.

Figure 4.47 Marbled White *Melanargia galathea*. Viewed from below with wings just past the halfway point of the downstroke.

Figure 4.48 Brimstone *Gonepteryx rhamni*. Viewed from below with wings just past the halfway point of the downstroke.

Figure 4.49 Green-veined White *Pieris napi*. Viewed from below with wings fully outstretched at the halfway point of the downstroke.

Figure 4.50 Brimstone *Gonepteryx rhamni*. Wings nearing their full down position towards the end of the downstroke.

Figure 4.51 Reconstructed take-off sequence in a Painted Lady *Vanessa cardui*. The insect is viewed from directly in front.

Figure 4.52 Orange-tip *Anthocaris cardamines*. Consecutive stages in the performance of a banked turn at take-off. The first downstroke (1) propels the insect backwards; the next upstroke and downstroke (2–5) throws the insect onto its right side and simultaneously turns its body 90 degrees to the right; the next upstroke (6–8) almost restores the insect back to the upright orientation and it exits at right angles to its original bearing.

Figure 4.53 Green-veined White *Pieris napi*. Consecutive stages in the performance of a banked turn. The insect is viewed from below. The first downstroke (1–2) and upstroke (3–4) turn the body through 90 degrees to its right; the next downstroke (5–6) turns it almost a further 90 degrees to its right. So, in less than two wingbeats the butterfly has turned through almost a full circle.

Arriving at the upper-stroke transition, the wings then peel apart in preparation for the downstroke. Again, these events are the same as those described in other insects. Stages in the peel and downstroke are shown in Figures 4.27–4.33 and 4.35–4.50 respectively, and from Figure 4.34 we see the similarity in airflow pattern during the upstroke and downstroke peeling processes. Figure 4.51 is a take-off sequence recorded using a half-second burst in a camera fitted with an extreme wide-angle lens.

All the features of the stroke that have been described can also be seen by careful inspection of the many high-speed videos of butterflies in flight. One clip in particular shows a male Orange-tip performing an exquisite, banked turn (HSV.20). Although the sequence lasts only a fraction of a second in real time, it speaks volumes about butterfly flight. The behaviour is summarised in Figure 4.52 using eight frames culled from the video. The insect launches backwards from a flower with a powerful downbeat of its wings, then in one continuous flowing movement rolls 90 degrees to its right, then 90 degrees to its left, then makes an exit at right angles to the direction in which it was originally facing. The entire manoeuvre took less than two wingbeats, and I needed to scrutinise the recording repeatedly before I could work out exactly what was happening in that fleeting three-hundredths of a second. In the panel shown in Figure 4.53 we see a Green-veined White butterfly do a similar banked turn in mid-air through nearly a full circle, again in less than two wingbeats.

Wing wear and tear

In springtime it is not unusual to come across butterflies with very badly tattered wings, which might seem surprising so early in the year until you recall that almost all of them – Peacocks, Red Admirals, Small Tortoiseshells, Commas – are individuals that emerged from pupation in the previous autumn before overwintering, and are now enjoying a new lease of life. So, the damage was inflicted many months ago and I always enjoy seeing these older specimens because their scars tell a story about their brushes with past perils. I understand why butterfly enthusiasts love to see fresh, picture-perfect examples. However, aside from good looks, their appearance tells you nothing about them as individuals. Some might jib at the idea of an individual insect possessing particularities that mark it out from other members of its species – surely all Red Admirals, say, are the same. Not necessarily so. Any animal that has survived to become 'old' by the standards of its peers has exceeded normal expectations, either by dint of good luck or, more likely, because it has drawn on its own resources. Wing damage, therefore, carries a narrative and there is a way of 'reading' it because it comes in two main forms. The first of these is ordinary wear and tear. Seeing the two bedraggled specimens resting their wings in Figure 4.54 you would be forgiven for thinking that they might never rise into the air again: however,

Figure 4.54 Small White *Pieris rapae*. Severe wear and tear of the wings. This kind of erosion occurs between the veins at the wing margins.

Figure 4.55 Small White *Pieris rapae*. Despite severe wear and tear at the edges, the main surface of the wings remains intact, and the insect is still able to fly as normal.

this would be to underestimate the resilience of butterflies (Figure 4.55). Saw-tooth erosion of this kind is the cumulative result of countless instances of wings snagging on thorny pieces of vegetation such as Teasel and thistle spines, especially during take-off (Figures 4.56–4.57). As the latter Figures make clear, elasticity allows the wing to bend without breaking, but it cannot prevent tearing. Erosion is limited to the areas between the veins, but the spoke-like veins themselves remain intact, projecting from the edge of the wing. One is reminded of an old, tattered umbrella that has seen many a stormy day and keeps out most of the rain.

The second kind of damage is a more obvious laceration often described as 'bird-strike'. The telltale sign of this is the absence of whole sections of wing, veins included – usually at the trailing edges. Figures 4.58 and 4.59 show an extreme case: an individual that had lost approximately one-third of the area of its hindwings and virtually the whole of its forewings, apart from their leading edges. It continued to forage on Lucerne flowers growing in a field, using an exaggerated flight style, and was careful to confine its activity to the border of the field within easy reach of a hedge in case of an emergency. A second example, a Large White, had lost a third of its left hindwing but was still able to perform a perfect 90-degree roll onto its side as it took off from a flower (HSV.21a). Figure 4.60 shows the flight sequence of a wing-damaged individual recorded during a half-second burst in a camera fitted with a wide-angle lens.

Cases such as these prove how effectively butterflies can compensate for injuries to their wings. Even when there is little left of the forewings except the toughened spar at the leading edge, function remains. We can infer from these observations that a fracture of the leading-edge spar of the forewing

Figure 4.56 Brimstone *Gonepteryx rhamni*. Wings chafing against spiny bracts of a Teasel flower at take-off.

Figure 4.57 Marbled White *Melanargia galathea*. Acute bending of the wing tip during take-off.

Figure 4.58 Small White *Pieris rapae*. Severe damage to the wings due to bird-strike.

Figure 4.59 Small White *Pieris rapae*. Severe damage to the wings due to bird-strike.

Figure 4.60 Painted Lady *Vanessa cardui*. Reconstructed flight sequence.

completely disables the insect and is in effect a death sentence, because it can no longer fly and suffers an inevitable fate on the ground. The only butterflies I have ever come across with a broken leading-edge spar were lying on the ground after being struck by passing cyclists or cars.

In terms of tolerance to damage or impairment, the hindwings are much less important. The Orange-tip shown in HSV.21b had an unusually hampered-looking flight style that was obvious even at a distance. Its hindwings had failed to expand during emergence from the pupal case and now projected from the body in the form of two pleated lumps that impeded the motion of the forewings.

In science, it is often many years before questions become regarded as having been finally 'settled': this applies to big subjects like the origin of the universe, the rotation of the planets around the sun and the circulation of blood around the human body, as well as little subjects such as the one

I introduce here concerning the role of eyespots in butterfly wings: Are they there to deter birds or to attract them? The natural enemy of butterflies in flight is assumed to be birds, and the popular view is that eyespots act as a target, distracting attention away from the vulnerable body. If this is the case, it contrasts with the presumed role of eyespots in the wings of the Iris mantis discussed in Chapter 1, which was to deter. The image of a flying owl butterfly *Caligo* sp. looking remarkably like a Long-eared Owl *Asio otus* (Figure 4.61) could be seen as supporting the deterrence argument and of course it is possible that eyespots serve different roles in different species.

In Britain, the species with the most prominent eyespots is the Peacock *Aglais io*, with four, one on each wing (Figure 4.62). Healthy-looking Peacocks are often seen with large pieces of their wings missing (Figure 4.63), so it could be argued that they have survived an attack that otherwise would have been directed at the body and killed them; this is compatible with the distraction theory. But it cannot be regarded as conclusive. First, it is equally feasible that birds strike indiscriminately at all parts of the butterfly's body and wings, including the eyespots. Those insects that were targeted specifically on the body would perish and seeing survivors with tattered wings would tell you nothing to the contrary. Second, species with very

Figure 4.61 Owl butterfly *Caligo* sp. Viewed from behind, a flying owl butterfly appears remarkably like a Long-eared Owl with a rather menacing look in its eye!

Figure 4.62 Peacock *Aglais io*. The eyespot on the forewing is situated on the outer section of the leading-edge spar. If it were targeted by a predator, the consequences for flight could be catastrophic.

Figure 4.63 Peacock *Aglais io* in flight with tear (bird-strike?) in the eyespot of the left forewing.

plain wings, such as Whites, show at least as much evidence of bird-strike as those with pictured wings.

Most bird-strike is located towards the trailing edge of the hindwings. This supports the notion of birds striking from behind while in pursuit. But this argument, too, is questionable: it is not possible to know how often the forewings may also have been targeted, because the victims of such attacks may not live to tell the tale, for the reasons mentioned above. In any case, the eyespots on the forewings of the Peacock are located just behind the leading edge, a particularly vulnerable area, as just described. Figure 4.63 shows a Peacock with a tear in the eyespot of the front left wing but also a tear on the right hindwing, ahead of the eyespot. I feel that the jury is still out on this question and a statistical study of the distribution of strike sites over the whole surface of the wing, including any eyespots, would be needed to solve the debate.

Occasionally butterflies are seen with a perfectly symmetrical pattern of bird-strike on the left and right wings. In theory, this could be the result of a bird snapping at the wings of a butterfly in flight at the precise moment when they clap together at the top of the stroke. But the tiny dwell time of the wings in this configuration argues against that idea and it seems more likely that the damage is inflicted when the butterfly is roosting with its wings closed (resting butterflies close their wings vertically above the body), either by a bird or a small mammal.

5 Flight Behaviour

Take-off technique

The origin of flight in insects is a mystery, not least because the geological record lacks any fossil equivalent of *Archaeopteryx*. One would expect such a fossil to show outgrowths of the thorax, abdomen or legs that could be interpreted as primordial wings. According to one hypothesis, wings originated as safety devices or parachutes for insects accidentally falling from trees and bushes. This seems a fanciful idea but shifts the spotlight away from the anatomical precursors of wings to practical methods for getting into the air in the first place, it being unlikely that the putative primordial wing could flap. The two options are: jumping from the ground or jumping/free-falling from trees. Free-fall automatically provides the pump-priming airflow around the wings; in fact, it is the normal way modern hawker dragonflies take off from a hanging position in trees and bushes. Curiously, evidence relating to the free-fall origin of flight comes not from winged insects, but from a primitively wingless group comprising the springtails, bristletails and two-tailed bristletails, the first two of which have specialised jumping mechanisms. Springtail jumping is based on a 'pogo-stick' model: release of a spring-loaded 'tail' hurls the insect through the air (Figure 5.1) but there

Figure 5.1 Springtail *Tomocerus* sp. The insect is jumping from left to right. Jumping results from the sudden extension of a fork-like 'tail' which is normally held in a groove on the underside of the abdomen.

is no anatomical equivalent of a shelf-like outgrowth from the body which could effectively convert the jump into a glide. Bristletails jump by suddenly curling the body downwards and in so doing lashing the ground with their feet and cerci – the three long bristles at the end of the body. Experiments have demonstrated that the cerci of tree-dwelling species can be deployed like rudders to influence the gliding descent path during free-fall. But they too lack any structure even remotely suggestive of a wing.

The arguably more likely method for setting up airflow around a static wing is jumping. Even if the ancestral wings were simple plate-like outgrowths, with no muscular provision for flapping, jumping would allow the insect to glide over short distances. An examination of the take-off techniques used by extant species of insects could offer insights into the origins question.

To casual observation, jumping is no more than a crude explosion of unguided activity with no other purpose than to propel an insect out of harm's way at maximum speed. Certainly, that could be the impression given as one tramps through rough grassland, sending up clouds of grasshoppers in every direction (Figure 5.2). Pausing for thought, you might then reason no, not every direction, they all jumped 'away' from me. So, the jumping is not random. In fact, winged species of grasshoppers and crickets have three options at their disposal when they jump: wingless jumping, wing-assisted jumping and what could be termed 'delayed' wing-assisted jumping. In studies of animal behaviour, it is always easier to describe responses than motivations; nevertheless, some simple correlations emerge.

Figure 5.2 Blue-winged Grasshopper *Oedipoda caerulescens*. Leaping.

Figure 5.3 Field Grasshopper *Chorthippus brunneus*. Reconstructed sequence of a wingless jump. The dashed line marks the position of the head as the insect spins during its journey through the air.

When a grasshopper jumps without the aid of the wings it achieves a quick getaway but almost invariably at the expense of throwing the body into an uncontrollable head-over-heels spin, an example being given in Figure 5.3 and HSV.22. But if the wings are recruited simultaneously with the legs at take-off, spinning is avoided; if they open when the jump is already underway, spin is eliminated in mid-air (HSV.23). Wingless jumping is probably faster by a few milliseconds because it avoids any delay associated with opening the wings, but it rules out any chance of the insect landing on its feet. This can be countered at any point of the jump by activating the wings, which also extends the length and duration of the jump. The relationship between jumping and spinning has potential relevance to the flight origins question: might it be possible that the putative pioneers of insect flight initially developed shelf-like outgrowths from the body – in effect, fins – as anti-spin devices to improve the chances of landing on their feet after jumping? This would pave the way to the next step which would be the conversion of the static fin into a flapping wing.

So, far from being a simple knee-jerk reflex, the jump is a variable response that can be tailored to circumstances as perceived by the insect. In HSV.24 we see a Field Grasshopper *Chorthippus brunneus* preparing to jump. Clinging to a grass stem, it is facing away from the camera but can still see events unfolding thanks to its all-round vision. To avoid the camera it quickly circles around the stem to the opposite side, where it now has a perfect view of the approaching object. If it estimates that the intruding object has stopped, it will remain fixed rather than give away its position; since the camera continued to advance, the grasshopper chose to jump backwards directly away from it. The sequence of events is too fast for the naked eye to see but there is clearly an element of premeditation in the grasshopper's response, which the high-speed camera captures but the casual observer would never suspect was happening.

If disturbed, the Long-winged Conehead cricket *Conocephalus fuscus* usually begins its jump wingless then opens the wings mid-jump (HSV.25); on other occasions it chooses not to jump, but to extend its hind legs and assume a cryptic stance (HSV.26). The forewings of the Tree Cricket *Oecanthus pellucens* are transparent and their veins are modified into a sound-producing organ or 'harp'. Incidentally, the pure, bell-like sound soars into the night sky and inspired one writer to compare it with the sound that the moon would make if it could sing. As in the case of the Field Grasshopper above, the Tree Cricket has the option of wingless or, as in the example shown in Figure 5.4, wing-assisted jumping. A male European Dwarf Mantis *Ameles spallanziana* launches through the air without opening its wings (Figure 5.5), whereas the female has no other choice because she is wingless.

Figure 5.4 Tree Cricket *Oecanthus pellucens.*

Figure 5.5 European Dwarf Mantis *Ameles spallanziana*.

Use of the legs to provide push-off from the ground is not restricted to specialist leaping insects, but forms part of the normal launch process in most flying insects – even the Housefly (Figure 5.6). Birds do the same thing; in fact, in their case an extra boost is given to the momentum by flexing the ankle just before leg extension. Human high-jumpers use a similar technique; in both cases the extra energy stored in the elastic Achilles tendons is released as the legs extend. In the next section of the chapter, we meet insects deploying elastic springs to achieve take-off accelerations hundreds of times greater than those of Olympic high jumpers.

Take-off in beetles can sometimes seem ponderous due to the extra time and care required, first to raise the elytra and then to iron out the various folds in the wings. The process is accompanied by exaggerated tip-toeing movements of the feet as the insect strives to attain perfect balance before taking the final plunge. With the preliminaries completed, it 'dives' into the air, all legs akimbo (Figures 1.3a, 1.4a, 1.26, 5.7, HSV.27). The routine is highly stereotyped, leaving little room for adjustment mid-sequence, and one sees ladybirds making false starts time and time again.

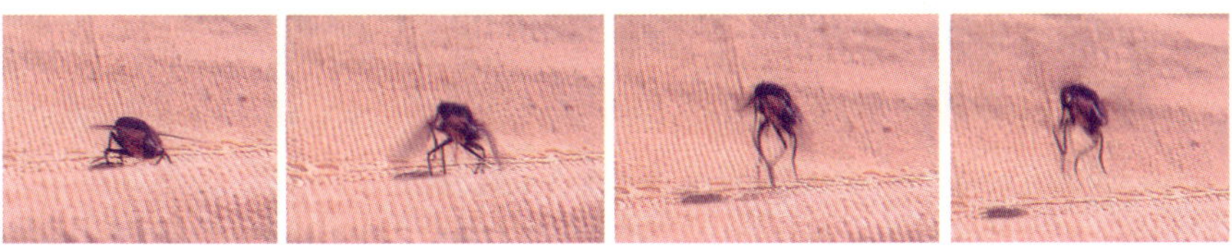

Figure 5.6 Muscid fly. Leg extension during take-off.

Figure 5.7 Black-headed Cardinal Beetle *Pyrochroa coccinea*. Consecutive stages during take-off.

Figure 5.8 Tiger beetle *Cicindela* sp.

Figure 5.9 Green Tiger Beetle *Cicindela campestris* in flight.

Tiger beetles lie at the other extreme, probably boasting the fastest wing opening of any family of beetles, a feature closely tied to their predatory habit. They use their long legs to run down prey over open ground, reserving their wings for an emergency escape whenever danger presents itself from birds or small mammals (Figures 5.8–5.9, HSV.28–30). Their exposed lifestyle therefore demands exceptional wing-opening speed: I have clocked the take-off time of the Green Tiger Beetle *Cicindela campestris* (wing opening plus leg extension) at approximately one-hundredth of a second – not as fast as some flies, for example, but probably faster than any other insect of a similar body size. Willingness to use the wings is highly dependent on temperature and a sudden dip in ambient light as a cloud passes in front of the sun seems to instantly paralyse the flight muscles. When this happens, the insect chooses to remain motionless, relying on its camouflage colouring to avoid detection.

When I attempted to record their behaviour on camera, I quickly discovered that they were far too alert for normal approaches to work. Getting close enough to record the behaviour of insects in the wild is always a challenge but easing a camera lens to within 4 or 5 cm of a tiger beetle demanded a different kind of patience – one could say 'tolerance'. Like all hunting insects, the tiger beetle is equipped with excellent all-round vision and, as I made my approach, it had already spotted me from several metres. A frustrating cat-and-mouse game then ensued, with me advancing a few centimetres and the beetle following suit, not yet sufficiently alarmed to make a dash for cover. If I managed to get within half a metre, the beetle seemed torn between curiosity and the need to escape. From that point onwards, it matched every step of my advance with a measured step of its own, centimetre for centimetre. By this stage, 'hunter' and 'quarry' were both moving at a snail's pace, the beetle (as it were) constantly looking over its shoulder, while I tried to remain as still as possible before taking the next step.

After many failures I realised that the 'lockstep' pattern of the beetle's movement meant that in its mind a threshold had been reached. In its daily life, it would be accustomed to potential threats speeding towards it and it would react quickly and instinctively. The slow approach confuses it; it cannot decide whether you are a shadow, a solid object or something vague in the sky. Hence its curiosity. Guessing this to be the case, I came to the conclusion that the only way to close the gap even further would be to physically shrink my silhouette by hunching forwards or bending backwards – both equally uncomfortable – making it appear to the insect that I was no longer looming down upon it but was moving away. Each of these many – perhaps hundreds – episodes was a minor battle of wills and my reading of the situation, being tinged with anthropocentrism, might have been very wide of the mark. But, from a purely practical point of view, in a contest like this the odds will always favour the native inhabitant because I am an intruder on its soil, and it can pre-empt my moves better than I can pre-empt its moves.

Butterflies are fond of basking in the sunshine with open wings and if one suddenly takes off you might expect that the very first movement would be to beat the wings down, since they are already halfway down anyway. But the butterfly does the opposite: with a flick it closes them above its body and then executes the first downstroke. In this way, it obtains the twin aerodynamic benefits of (a) the downstroke peeling process and (b) a full beat of its wings, rather than just half a beat. Further impetus is given to the take-off by active extension of the legs (Figures 5.10–5.11, HSV.31) although the contribution from such frail appendages must presumably be small. Species in the habit of hanging onto the sides of flowers to feed, such

Figure 5.10 Marbled White *Melanargia galathea*. Take-off sequence. 1 and 2 show leg extension synchronised with the first downstroke of the wings. 3 and 4 show the following upstroke.

Figure 5.11 Brimstone *Gonepteryx rhamni*. Leg extension synchronised with the first downstroke of the wings during take-off.

Figure 5.12 Brimstone *Gonepteryx rhamni*. Reverse take-off.

Figure 5.13 Large White *Pieris brassicae*. Reverse take-off.

Figure 5.14 Marbled White *Melanargia galathea*. Reverse take-off. The first downstroke has pitched the insect onto its back. It is now in the middle of the upstroke which will restore the body to the horizontal.

Figure 5.15a Backswimmer *Notonecta maculata* in flight. As in other Heteropteran bugs, most of the forewing is toughened, the rear section (embolium) overlapping the leading edge of the hindwing. The long, oar-like hind legs provide push-off from the ground during take-off.

Figure 5.15b Backswimmer *Notonecta maculata* in flight. Wings captured at the upper-stroke extreme (left) and lower-stroke extreme (right).

as Brimstone and Whites, take off backwards (Figures 5.12–5.13), and in an emergency this forms the basis of a spectacular high-speed manoeuvre in which the first downstroke flips the insect onto its back (Figure 5.14) and the following upstroke pitches it back to the horizontal – the entire 'flip-flop' action taking only a few hundredths of a second (HSV.32).

The pond-dwelling backswimmer *Notonecta maculata* provides one of the most unexpected cases of a jumping take-off (Figure 5.15a, b). Its long paddle-like hind legs are specialised for swimming, not terrestrial locomotion, but when occasion demands, such as the need to disperse in search of other backswimmers or new feeding grounds, it emerges onto the ground at the waterside, folds its legs beneath its body and leaps into the air.

Escape mechanisms

Jumping insects such as grasshoppers, crickets, fleas, flea beetles, froghoppers and leafhoppers are familiar to most people, but there are many more species whose jumping abilities have yet to be recorded. The Alder Fly *Sialis lutaria* is a common sight on the riverside in spring as it flits restlessly from leaf to branch to grass stem. From casual observation over the years, I had formed the impression of a brief, vague 'hiccup' in its motion as it took off. The high-speed camera showed what is happening. During a normal take-off the insect uses its legs to push off and synchronises this with the opening of the wings (Figure 5.16), but a more interesting situation unfolds when it leaps without opening the wings; then it performs a series of forward somersaults (Figures 5.17–5.18; HSV.33). This high-speed escape mechanism is the 'hiccup' that had previously confused me and presumably has the same effect on birds and other would-be predators.

A similar behaviour is exhibited by another riverside inhabitant, the caddisfly. Like the Alder Fly, it is an unlikely looking high-jump candidate, but in Figures 5.19 and 5.20 and HSV.34a–d we see it using its middle and hind legs to leap into the air. As with many other insects, the leap is a normal part of the take-off routine (Figure 5.21), but if an emergency requires it

Figure 5.16 Alder Fly *Sialis lutaria*. Reverse take-off.

Figure 5.17 Alder Fly *Sialis lutaria*. Consecutive stages in the forward somersault escape reaction.

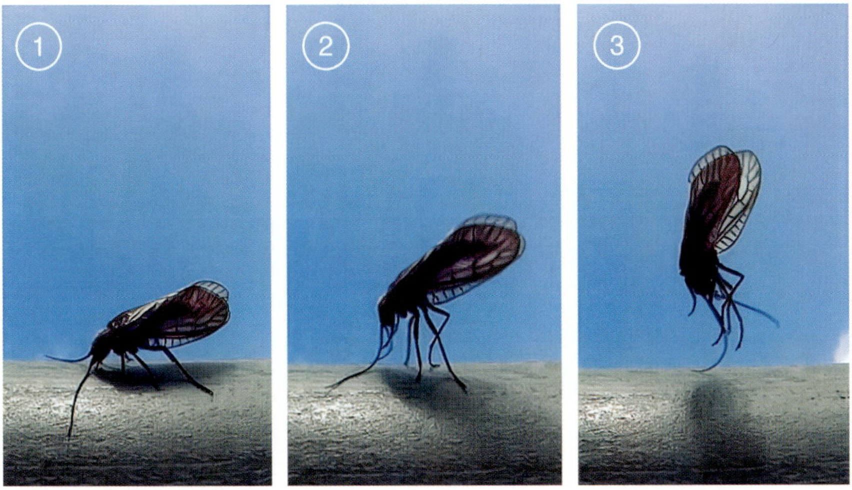

Figure 5.18 Alder Fly *Sialis lutaria*. Leg extension during forward somersault.

Figure 5.19 Caddisfly. Consecutive stages in a somersaulting jump. This is a rapid escape reaction and there is no attempt to risk delay by opening the wings.

Figure 5.20 Caddisfly. Consecutive stages in a leaping take-off into flight. This insect is leaping into flight; there is no body rotation and by stage 3 the wings are beginning to open.

Figure 5.21 As the sun begins to set over an East Anglian fenland ditch, a caddisfly launches into flight.

the caddisfly dispenses with wing opening and the jump becomes a rapid escape response. When I first recorded this behaviour I thought I was the only person ever to have seen it, although I later found that Malcolm Burrows, a former professor in the department of Zoology at the University of Cambridge, had carried out a laboratory study on the very same subject many years earlier.

The small fry of the insect jumping community include froghoppers, leafhoppers and flea beetles, and at first sight their behaviour has no more method than sparks flying from a fire. In one sense this is true. The jumping mechanism is based on the explosive release of energy stored in a spring and, per unit of body weight, it packs far more punch than any of the insects mentioned so far. Since the packet of energy is released in an all-or-nothing manner, the insect has no control over its speed of escape, but it can control range and direction by synchronising the action of the legs and the wings. On different occasions a flea beetle might jump either with or without opening

Figure 5.22 Flea beetle. Wingless jumping with body spin.

its wings (HSV.35–36). As in the caddisfly, Alder Fly and grasshoppers, wingless jumping produces a faster exit but does so at the expense of sending the body into a high-speed spin, meaning that the insect is very unlikely land on its feet (Figure 5.22). Spin can be avoided if the legs and wings are deployed together at take-off, and in this case a feet-first landing is almost guaranteed (Figures 5.23–5.24). Surprisingly, it appears that these apparently very 'simple' insects have at their disposal mechanisms to tailor their escape to different circumstances, rather than being slaves to a single invariant response.

To investigate this question further, I conducted a very simple study on the flea beetle *Aphthona* sp. When I presented the insects with a brightly lit target, they invariably turned to face it, set their feet in a very precise manner as though aiming at the target, then leapt towards it. As expected, those that jumped without opening the wings spun rapidly through the air and virtually never hit the target feet-first, whereas those that recruited

Figure 5.23 Flea beetle. Wing-assisted jumping without body spin.

Figure 5.24 Flea beetle *Phyllotreta* sp. jumping with wing assistance. The wings have been captured near the end of the downstroke. Wing assistance prevents body spin and virtually ensures a feet-first landing. Note that as in beetles in general, the wing covers or elytra make little active contribution to lift production.

the wings almost always did. In both cases, I also measured the accuracy of their targeting performance. For an insect possessing a compound eye with no more than 60 pixels (compared, for example, with a dragonfly's 30,000), the results were impressive. The hits were focused not on the actual target centre but at a point directly below on a line passing vertically through the centre of the target (Figure 5.25). This is a result you might expect in a darts player who had perfect vision and a steady hand yet stubbornly kept hitting the board at a point just below the bull's eye. Why? Because they failed to consider the fall of the dart due to gravity. From the performance of the flea beetle, we can conclude that its 'line of sight' is remarkably good but, as would be expected, its brain is not wired to deal with ballistics. It would be unwise to compare the results of what is in effect a laboratory study with a potential situation in the wild, however. What does a brightly lit target

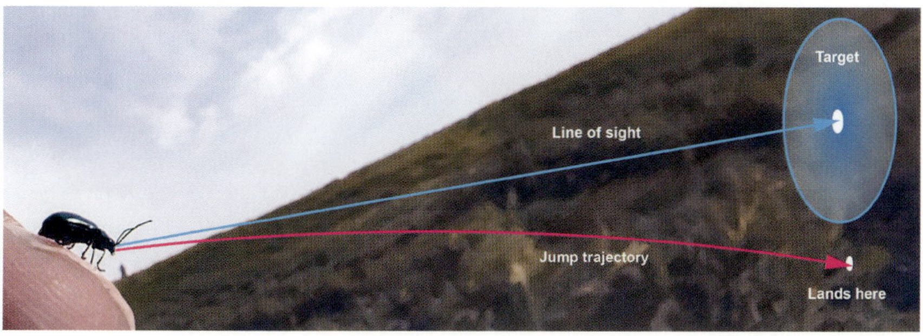

Figure 5.25 Flea beetle. Jumping trajectory towards an illuminated target. The insect sets a trajectory along the line of sight towards the target centre, but during the jump gravity pulls its body down so that it intercepts the target below the centre.

mean to a flea beetle? Sunlight shining through a gap in foliage, offering a potential escape route from a stressful situation? Or were the insects simply transfixed by the bright source of light and felt compelled to jump towards it? Even if the latter were the case, the results at least demonstrate remarkable powers of perception and eye/muscle coordination.

Froghoppers and leafhoppers (Figure 5.26a) possess the most powerful springs of all, and some species can attain take-off accelerations of several hundred g (the acceleration due to gravity, 10 m per second per second). To put that into perspective, a trained astronaut on their way to space in a rocket will rarely experience more than 5 g and could not possibly survive 10 g. The froghopper survives because its 'high g' experience lasts for only a few thousandths of a second. In HSV.37a pair of red-and-black *Cercopis vulnerata* leap from my finger and it is interesting to note their behaviour as they prepare to do it. The normal mating position (a subject dealt with in greater detail in the next chapter) of froghoppers is back-to-back. Before jumping, mating *Cercopis* realign their bodies so that they become almost parallel and both insects now face in the same direction. Then they jump and after a fraction of a second in the air open their wings and fly, still coupled together. The body adjustments are carried out with care and are

Figure 5.26a A leafhopper jumping, with assistance from the wings. Activating the wings prevents body rotation and may extend the length and duration of the jump.

Figure 5.26b Great Green Bush-cricket *Tettigonia viridissima*.

understandable when we consider the forces involved in the jump. The acceleration of a related species, the Common Froghopper *Philaenus spumarius*, has been timed at 400 g; this means that during the brief time that its legs are extending, the froghopper must be exerting a force equivalent to 400 times its own body weight. This poses the question: could such a force damage the abdominal organs of a mating pair, if they jumped facing away from each other in the normal back-to-back mode? For example, seen in human terms, the combined pull between the two bodies leaping together would be equivalent to a person weighing 70 kg hanging from a horizontal bar with a weight of 56 metric tons attached to their ankles – albeit for only a millisecond or so! Likewise, the tensile stress on the insects' bodies would be felt for only a fleeting moment of time, but their actual pre-jump behaviour implies that the risk is something to be avoided.

Virtually all the jumping insects that use spring propulsion lie at the very small end of the size spectrum, whereas those that rely on real-time muscle action include some of the biggest of all insects. Why is this the case? To understand it, let us compare the jumping performance of two insects lying at the opposite extremes: a tiny leaf hopper and the Great Green Bush-cricket *Tettigonia viridissima*, one of Europe's biggest insects (Figures 5.26a and b). *Tettigonia* is seen leaping into action in HSV.37b, incidentally also producing an impressive display of hindwing peeling once flight is underway. It can be shown using elementary mechanics that the height of a jump is strictly related to the take-off velocity regardless of the size of the insect involved. To jump vertically through a height of 0.1 m, for example, they must both achieve a take-off velocity of 1.4 m per second, equivalent to a quick walking pace by human standards. The important variable here is the length of the hind leg when fully extended at take-off. Very roughly, let's suppose this is 1 mm (0.001 m) in the leaf hopper and 5 cm (0.05 m) in the bush-cricket. We can now calculate the time that a leg must be extending to achieve a take-off velocity of 1.4 m per second; this is given by (leg length/take-off velocity). The answer is 35 milliseconds in the bush-cricket, compared with 0.7 milliseconds in the leaf hopper. The former falls comfortably within the range of normal muscle contraction speeds; the latter – at less than one thousandth of a second – is perhaps fifty times faster. So, to achieve jumping speeds and heights that are of practical value in escaping from potential enemies, very small insects are forced to rely on mechanical springs.

6 Male Flight Behaviour

Courtship and rivalry

The general subject of insect behaviour is outside the scope of this book, which is concerned mainly with specific instances related to flight, and particularly the behaviour of males during courtship, swarming and dancing. The dominating theme is rivalry, sometimes expressed in the form of elaborate rituals but often simply involving two or three males vying for the attention of a female, as shown in Figure 6.1 and HSV.38, where two male droneflies *Eristalis pertinax* hover so close to a feeding female that she can probably feel the downwash from their wings, even though she shows every sign of indifference to their presence. You always know when such an encounter is taking place nearby because the droning – which in this species sounds more like a 'whine' – is twice as loud, as in the combined sound of two propelled aircraft passing overhead in close formation.

As an example of a highly muddled encounter between male and female of the species, the courtship behaviour of members of the sawfly genus *Cephus* would be hard to beat (Figure 6.2, HSV.39). In early spring, clusters of males and females cram themselves into single buttercup flowers in a knotted mass of wings and legs. In Charles Darwin's classic theory of sexual selection, males assert themselves through either combat or display, but neither of those descriptions fits the behaviour of the sawfly and precisely which male succeeds in winning the female seems more like a lottery. The first thought that enters my mind when I see this scrimmage is not some deeply challenging question pertaining to selfish genes but 'How did all these insects meet in the first place: was it by chance or by design?' *Cephus* is drawn to yellow flowers, especially buttercups, so inevitably there will be many incidental encounters during foraging. More generally, however, inter-species recognition and attraction between the sexes depend on definite cues, usually in the form of chemical and visual signalling. Wind-borne female hormones (pheromones) exert their effect over relatively large distances, while visual clues become more important at close range.

The colours and markings on the bodies of some insects are so striking that one instinctively thinks they must have some kind of appeal, either to the owners themselves or to fellow members of the species. This reveals the bias of human thinking, because, as we saw in Chapter 1, often they are meant for other eyes, namely those of their chief predators, birds. On the same theme, in Chapter 3 we saw that the ground colour of the wings, rather than the pattern of markings on them, is the dominant feature in

Figure 6.1 Dronefly *Eristalis pertinax*. (1) a feeding female attended by two hovering males; (2) the female flies off, and the males begin their pursuit; (3) the female reappears at top left of frame.

Figure 6.2 Sawflies *Cephus* sp. Male and female encounter with upturned abdomens and waving wings.

interspecific recognition between butterflies and examples were given of the confusion this can create in males of the various species of Whites. In Figure 6.3 and HSV.40 a Large White swoops down to investigate a male Orange-tip feeding on a flower, apparently mistaking it for another Large White and evidently paying little attention to the splash of bright orange at the tip of the forewing. Even when a male makes the right choice, he may be rejected by the female because she has already mated. She signals this by adopting the well-known posture with upturned abdomen (Figures 6.4–6.5). Persistent individuals often ignore the sign and aggressive raids by rogue males on mating pairs are common. HSV.41 shows a male, Green-veined White harassing a mating pair on the ground and continuing to hound them in the air as they attempt to escape. A simpler way of flagging rejection is wing-waving, as seen in this encounter between two Clouded Yellow butterflies (Figure 6.6, HSV.42) and, in the next chapter, where a female

Figure 6.3 Large White *Pieris brassicae* and Orange-tip *Anthocaris cardamines*. A Large White appears on the scene, descends to investigate the Orange-tip, then flies away.

Figure 6.4 Green-veined Whites *Pieris napi*. The male flutters around the female as she signals that she has already mated by raising her abdomen.

dragonfly in the act of egg-laying is hustled out of the water by a male of another species (HSV.43).

I think when we put nature under the microscope, we should not rule out the possibility that on some occasions, even in insects, the animal may experience pleasure in what it is doing. To an onlooker, a word that could easily spring to mind when watching a ribbon of half a dozen butterflies twirling through the air in a game of chase would be 'glee'. If indeed it is rivalry, then it seems remarkably good-natured. By the same token, the two Marbled Whites seen in HSV.44 dancing around one another look as though they are just revelling in their own acrobatic abilities.

It is interesting that some of the most gregarious butterflies, such as Marbled White, Gatekeeper *Pyronia tithonus* and Meadow Brown *Maniola jurtina*, show little tendency towards organised territorial behaviour, whereas their rather more solitary relatives Red Admiral, Comma *Polygonia c-album*

Figure 6.5 Green-veined White *Pieris napi*. Female rejection behaviour. A male pesters a female even though she is signalling that she has already mated by raising her abdomen.

and Painted Lady *Vanessa cardui* are very zealous in the defence of territory. An east–west track running across open countryside is a favourite location, because it receives warm sunshine throughout the day but especially in the evening, when, as the air temperature begins to drop, the insects can spread their wings and put their back to the sun. Such locations present ideal opportunities for photographers wishing to capture butterflies rising into the air, wings backlit by the sun (Figure 6.7) – revealing, incidentally, the surprising translucency of butterfly wings.

The Speckled Wood *Pararge aegeria* is one of the least social butterfly species, preferring to set up station in a quiet sunlit woodland glade (HSV.45), jealously guarding its patch, seeing off every intruder and often returning to the same favourite leaf after each sortie. As an instance of determined territorial defence, however, I have seen few more impressive than the one featuring an aged Red Admiral in HSV.46. The location was

Figure 6.6 Clouded Yellows *Colias croceus*. The insect sitting in the foreground signals rejection of the intruder by opening and closing its wings.

Figure 6.7 Painted Lady *Vanessa cardui* rising in late evening to defend its territory on an east–west-running farm track.

a track shared and patrolled by four or five other much younger males, each laying claim to a stretch of about 10–20 m of ground. Any passing butterfly, bird, mammal or human walker would be challenged from the air. In contrast with the junior, fresh-liveried males, the older butterfly – probably a survivor from the previous year – had worn, tattered wings. Nevertheless, it was as determined to hold ground as any of the others, vying with them regularly in flight and even bold enough to land on my hand and allow me to film it.

Hovering and dancing

Normally the term 'social insects' is used in connection with species living permanently in communal groups, such as termites, ants, social wasps and bees; but social behaviour of a more seasonal kind is common in insects, examples being the migration of Monarch butterflies, the swarming of locusts and ladybirds and the aerial congregation of Cockchafer beetles *Melolontha melolontha* following a mass emergence from their pupal cases.

Figure 6.8 Winter gnats *Trichocera* sp. Performing a swarming dance.

Figure 6.9 Parasitic wasps swarming around a flowerhead.

Swarming Face Flies *Musca autumnalis* (close relatives of the Housefly) are less familiar (HSV.47), but most people cannot fail to have noticed winter gnats (Trichoceridae: not actually gnats, but small craneflies) dancing in the sunshine on mild sunny days in December or January. Despite their flimsy appearance, winter gnats are among the hardiest of British insects, rarely if ever feeding but always ready to forsake a hiding place in the shadows for a spot of dancing in the sunshine. As a physical phenomenon, dancing in insects is just like dancing in people: it is a choreographed display of balance and rhythm. Naturally the 'steps' of the insect dance are simpler, and the timing is not as precise, but the psychological effect is similar: to enthral both dancer and spectator via the medium of rhythm. The attitude of the body and legs of the winter gnat during the dance is characteristic, all six long, spindly legs splayed like the spokes of a cartwheel (Figure 6.8). This arrangement looks pleasingly decorative to the human eye but is in fact a highly functional device. Due to their drag on the air, the legs function like parachutes or aerial brakes, reducing the sinking speed of the body. The same principle can be seen in clouds of tiny parasitic wasps dancing around a flowerhead (Figures 6.9–6.10, HSV.48).

Figure 6.10 Parasitic wasps swarming.

Figure 6.11 St Mark's Fly *Bibio marci*. Male, performing hovering dance. (1) wings in the end-downstroke position; (2) wings in the end – upstroke position. The curved structure above the left wing is the left front leg bent back over the head and thorax.

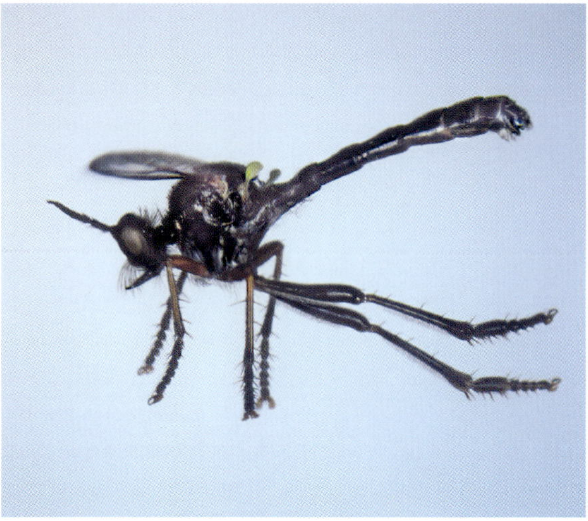

Figure 6.12 Robber fly (Asilidae). The long, trailing legs of insects such as this can be used to adjust the balance of the body during different flight manoeuvres. Here they are held at an angle that reflects the angle of the abdomen.

Male communal dancing can be viewed from a biomechanical point of view as 'quasi-hovering'. True stationary hovering, as we saw in Chapter 1, is energy-intensive and only a few specialist insects can sustain it for more than a fraction of second. A cheaper alternative, which still allows the insects to remain in a close grouping, is to move up and down or weave from side to side. In springtime, recently emerged male St Mark's Flies *Bibio marci* congregate in loose formations to dance in the shade of flowering trees and bushes, long hind legs trailing at an angle to the rear, thorax tilted upwards (Figure 6.11, HSV.49). The arrangement closely echoes that seen earlier in the hovering Red-banded Sand Wasp *Ammophila sabulosa* (Figure 2.2) and the Digger Wasp *Sceliphron destillatorium* (Figure 1.26), as well as slow-flying beetles such as the Wasp Beetle (Figure 1.25) and the Common Soldier Beetle (Figure 1.3a), and mirrors almost identically the attitude of the body in a hovering robber fly (Figure 6.12). The upward tilt of the thorax brings the stroke plane of the wings closer to the horizontal, which directly benefits the hovering process, and any consequent imbalances seem to be countered by precise repositioning of the hind legs. In the St Mark's Fly, the balancing act is completed by swinging the front legs back over the head and thorax, where they could easily be mistaken for long, curved antennae (HSV.50), a curious configuration that I have not seen in any other flying insect. The slow-motion video (HSV.49) shows the entire body vibrating in harmony with the wings, right down to the tip of the tarsi.

On still evenings in springtime, clouds of midges of the non-biting variety (Chironomidae) rise from the undergrowth to begin their ritual dance (Figures 6.13–6.14). Often you first notice them if you happen to look up and see a column of these tiny insects – no more than 2–3 mm in length – riding in the thermal coming from the top of your head. You then cast your

Figure 6.13 Chironomid midges swarming at sunset. There was a slight wind from the left and most of the insects have orientated into it.

Figure 6.14 Chironomid midge. Stages in the wingbeat cycle. The stroke angle is small: (1) supination at the upstroke extreme; (2) halfway through the downstroke; (3) supination at the downstroke extreme.

Figure 6.15 Chironomid midges. The streak photograph illustrates the sideways motion of members of a swarm hovering in a wind coming from the left.

eye around and become aware of countless other, much bigger, columns rising and falling rhythmically in the low sunshine. Each of the columns is a discrete entity separated from its neighbours by only a short distance but with little sign of any toing and froing of individuals from one column to the other. Inside the column individuals seem to move randomly, except that they respect its boundaries. If the stillness of the air is broken by a light breeze, the behaviour of the column changes and a more organised pattern of dancing emerges. Now the insects all face directly into the wind and begin a rhythmical backward and forward movement (HSV.51) that is captured well in streak photography (Figure 6.15, HSV.52). It seems that the wind gives the swarm the directional cue that it needs to graduate to a more synchronised form of dance; and the stronger the wind the faster the rhythm of their movements. At the same time, the collective whining of their wings grows louder and adds to the overall hypnotic effect.

Pivotal to this display is the fact that the midges appear simply to be following their instinctive understanding of the wind: by orientating into the flow, they gain a free ride as they fall backwards, then free lift under their wings as they drive forwards. By this means, an insect that is incapable of stationary hovering achieves the same objective by flying backwards and forwards over a measured distance. The stroke angle is quite small – I would estimate no more than 45 degrees. Figure 6.14 shows positions of the wings at three stages of the stroke.

Standing in open countryside surrounded in all directions by clouds of gnats, it is impossible not to think of the sheer numbers involved. A very crude estimate based on the density of columns and the average number of individuals within each column suggests a figure of about a billion per square kilometre. In an area of five to ten square kilometres there could be as many chironomid midges in the air as there are people alive on the Earth today: about eight billion. This would represent a minuscule fraction of the worldwide population of this single species. One is given pause to reflect not only on the almost unimaginable abundance of insects but also the vast cost in biological resources required to maintain the continuation of even a single species. What percentage of this army of males vying for the attention of a female succeeds in mating with one? The life of a male chironomid midge revolves entirely around his dancing; he has no other purpose. Yet I doubt if even one in a hundred fulfils his mission. His overwhelmingly likely fate is illustrated in Figure 6.16: he becomes spider fodder. A spider's web is in effect an almost perfect filter for any small insect whose journey through

Figure 6.16 Midges trapped in a spider's web.

Figure 6.17 Mayfly. Arrows indicate rising and falling phases of the dance. Most of the descending phase is passive free-fall with open wings and spread tail bristles. Power is applied towards the bottom of the falling phase and continues to the top of the rising phase.

the air happens to coincide with that window in space – with a stress on 'almost'. Watch HSV.53 and pay attention to the left-hand side of the frame, where you will see a midge come into view and fly straight through the web, jostling a couple of less fortunate individuals on the way through.

In the final three examples the insects dance up and down with a rhythmical bobbing motion. The rising phase is powered by the wings, the sinking phase is a free-fall through space, so in effect 'bobbing', as one might call it, routine is a midge-style dance in the vertical rather than the horizontal direction. The insects in question are the mayfly, the longhorn moth *Adela reaumurella* and the Common Blue butterfly *Polyommatus icarus*. Figure 6.17 reconstructs a single cycle of the mayfly dance from HSV.54, with the falling phase on the left and the rising phase on the right. During the falling phase the insect sinks passively through the air, with the wings and tails fully extended to achieve maximum air resistance and aerial breaking. The wings then resume activity, decelerate the body and power it back up to the starting point.

The dance of the male longhorn moth follows a similar pattern (Figure 6.18 and HSV.55). In this case the hugely exaggerated antennae of the male (those of the female are much shorter) and the long hairs clothing its body play the role of aerial brakes. Braking extends the duration of the fall, and this effect, together with the repeated nature of the dance components, presumably catches the eye of nearby females as well as demonstrating flight prowess to other males.

All these oscillating dances could be described – to continue the human analogy – as 'quick time', each cyclical 'step' lasting less than a second. A high-speed camera is needed to capture events for the human eye, but in the insect's own time frame they might well seem quite stately. The encounters are very brief – the example shown in Figure 6.19 taken from HSV.56 lasted just under one second, but in that fleeting moment five individuals met, interacted in a choreographed way, then parted company, probably never to see one other again. The scenario opens with a single individual beginning a bobbing routine that, typically of butterflies, compresses into a single stroke of the wings an activity that in other insects would require multiple strokes. The fall phase is in effect a slow version of clap and peel,

Figure 6.18 Longhorn moth *Adela reaumurella*. Stages of the wingbeat cycle during dance. (1, 3) end downstroke; (2, 4) end upstroke.

Figure 6.19 Common Blue butterfly *Polyommatus icarus*. Beginning and end phases of a dance. Butterfly (a) begins its bobbing routine and is quickly joined by four other insects following suit.

the body initially sinking under gravity with wings closed above the body. As the fall progresses, the wings begin to separate and under power from the downstroke the body reverses and is driven back up to the starting position. This is how the dance begins; then four more butterflies in the immediate vicinity join in the throng.

7 Mating Flight

The only legacy that non-human animals bequeath to the world is a blueprint in their genes, an infinitesimally small parcel of chemical instructions with infinite genetic potential. Mating, together with the ensuing process of egg-laying, is the culminating act in the lives of insects. Since the physical encounter temporarily restricts the movement of the insects, it is a time of increased vulnerability to predators. Winged insects tend to inhabit open countryside and are therefore more exposed to attack than ground-hugging species. The question emerges: to what extent – if at all – do the members of a mating pair remain capable of using their wings in an emergency? The chief restriction on flight readiness is the mating position, the two usual variants being 'back-to-back' and 'piggy-back'. A second consideration is body weight: even if one member of the pair remains free to use its wings, is it strong enough to carry the other member with it? Back-to-back mating leaves both sexes with the freedom to use their wings, but the male and female are facing in opposite directions, so their efforts will tend to cancel.

Some insects, such as beetles and true bugs (Heteroptera), seem never to mate in flight, but it is a common occurrence among 'true' flies (Diptera), three examples of which are given here, all mating in the back-to-back position:

Figure 7.1 Dark-edged Bee-fly *Bombylius major*. Mating pair.

Figure 7.2 Craneflies. Mating pair.

a bee-fly (Figure 7.1), a cranefly (Figure 7.2, HSV.57) and the St Mark's Fly (Figure 7.3, HSV.58a–c). In all cases, both sexes of the pair actively use their wings but in the latter two the female is considerably bigger than the male and, as a result, she sets the direction of flight. The male has little control over events and appears to be nothing but a hindrance to his partner? In fact, by deploying his wings he is actively supporting his own weight in the air at no additional expense to the female.

The male dung fly mounts on the back of the female, pinning her wings with his legs, so that, if the need for flight arises, he has no option but to carry her (Figures 7.4–7.5, HSV.59a). Muscid flies such as the Greenbottle *Lucilia* sp. and allies adopt the same position but in this case the male sits further back, leaving the female free to use her wings (HSV.59b). The male dagger fly must deal with an additional problem. It is not uncommon to see him in springtime, hanging from a twig by two or sometimes just a single front leg with a female in his grasp below him, who is herself clutching an item of prey that he has presented to her as a gift (Figure 7.6, HSV.60). Her wings are usually pinned by his legs so in flight he supports a load of perhaps three times his own body weight. If the female happens to drop her prey and the male loosens his grip on her, she is free to use her own wings, and the pair usually separates in mid-air (HSV.60).

Butterflies mate in the back-to-back position, so the same problems arise as in the case of the St Mark's Fly and the cranefly. We saw earlier in HSV.41 a rogue male Green-veined White harassing a mating pair, and it was clear that when they were forced to try to escape only one of them was actively using its wings, the other trailing passively behind. In mating pairs of Green-veined White (Figures 7.7, 7.9, HSV.61), Gatekeeper (HSV.62), Meadow Brown (HSV.63) and Small White (Figure 7.8, HSV.64) only one member does the work of flying, but it is difficult to say which, because the

Figure 7.3 St. Mark's Flies *Bibio marci*. Mating pair, with the male on the right and the female on the left.

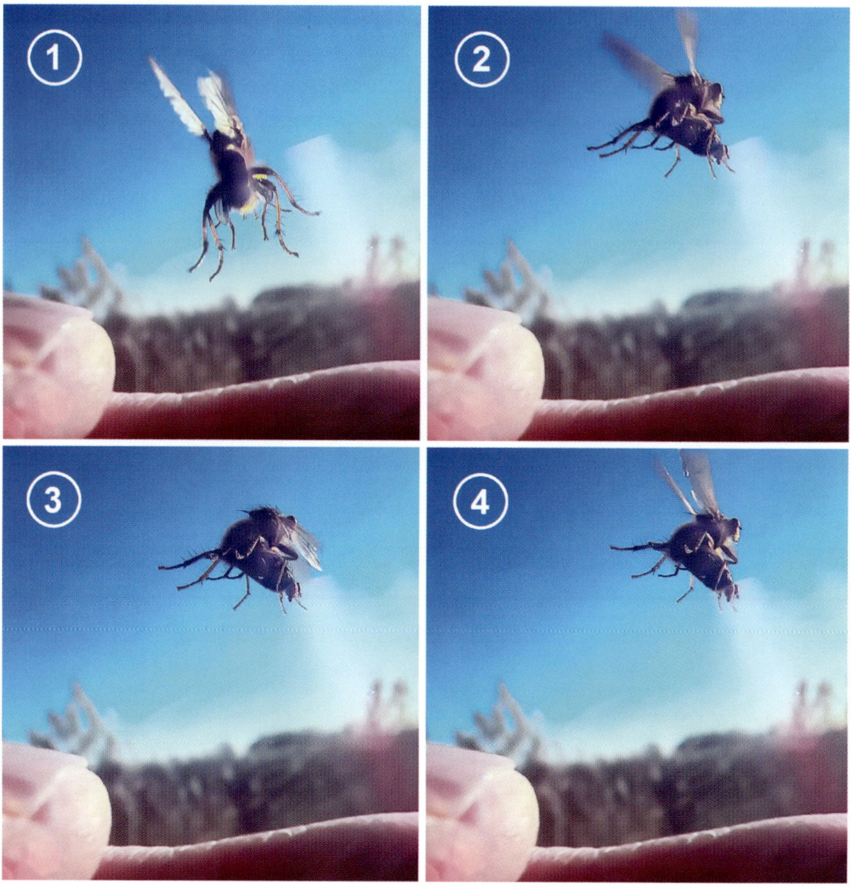

Figure 7.4 Yellow Dung Flies *Scathophaga stercoraria*. Mating pair. The female's wings are pinned by the male's legs. Consecutive stages in a single wingbeat of the male. (1) end – upstroke; (2) start of downstroke; (3) end-downstroke; (4) end-upstroke.

sexes look so similar in flight. Blues are an exception, since only the male is largely blue and the evidence is that he normally does the flying, the female hanging passively below him (Figures 7.10–7.11). On occasions the female takes a temporary lead at take-off (Figure 7.12), but the task of flying is quickly handed over to her partner. One assumes that the situation in the Common Blue is typical of butterflies in general and in principle this would represent an efficient division of labour between the sexes. The female has work lying ahead: she must bear the burden of producing eggs in her body, as well as expending energy in seeking out appropriate host plants on which to lay them. To observe a female Small White butterfly tracking down the last remaining sprig of Jack-by-the-Hedge *Alliaria petiolata* in the garden is a testimony to her doggedness, but also to the urgency that she feels. Time is running out, and her eggs need to be laid.

Figure 7.5 Yellow Dung Flies *Scathophaga stercoraria*. Mating pair. The female's wings are pinned by the male's front and middle legs.

Figure 7.6 Dagger flies *Empis tessellata*. Trio of male, female and prey item.

Figure 7.7 Green-veined White *Pieris napi*. Mating pair.

Figure 7.8 Small White *Pieris rapae*. Mating pair.

Figure 7.9 Green-veined White *Pieris napi*. Mating pair.

Figure 7.10 Common Blue *Polyommatus icarus*. Mating pair.

Figure 7.11 Common Blue *Polyommatus icarus*. Mating pair. The male is leading, the female trails below with open, motionless wings.

Figure 7.12 Common Blue *Polyommatus icarus*. Mating pair. In this case, the female temporarily takes the lead at take-off but soon surrenders it to the male.

The physical strain on the body of the male as he launches into the air with his mate is palpably evident in high-speed video clips. HSV.65 captures the contortions of his body as he forces his wings to their physical limits to deal with the awkward load. Once height and speed have been gained, his task becomes easier, and it is not unusual to see pairs eventually disappearing over the treetops.

There are two in-flight mating configurations in dragonflies and damselflies: first, the familiar 'copulation wheel', which is formed as the male grasps the female behind her head with the claspers at the end of his abdomen, while she curls her abdomen forward to make contact with the sperm store on the underside of his abdomen (Figures 7.13–7.16, HSV.66); and, second, the tandem formation that results when the female releases her hold on the male and is ostensibly pulled along by him, although the latter would be a misguided impression, as we shall soon see. Some dragonflies lay their eggs while in tandem: typically, the pair hovers above the selected

Figure 7.13 Damselfly copulation wheel.

Figure 7.14 Darter dragonfly copulation wheel.

spot, periodically dipping down towards the water, allowing the female to briefly immerse the tip of her abdomen below the surface to deposit an egg (HSV.67).

Both sexes take an equal share in the effort of flying and this is no surprise in insects for whom coupled flight is a normal part of their lifestyle and not a condition limited to emergencies. It seems logical, too, that natural selection would have ensured that mating flight is just as aerodynamically efficient as singleton flight. Matching speed for speed is a practical necessity because mating pairs rarely find themselves free of harassment by 'rogue' males. Acceleration from take-off in the dragonfly copulation wheel seems exceptionally fast, perhaps faster than that of uncoupled individuals, and there are clues as to what the mechanism behind this might be. Coincidentally, this subject takes us back to the problem of the extinction of the giant dragonflies of the Carboniferous era discussed in Chapter 1. According to one theory that has received relatively little attention, increasing body size led to the abandonment of the copulation wheel because the insects had become so large that the male could no longer carry the female. Presumably this then

Figure 7.15 Chaser dragonfly copulation wheel.

Figure 7.16 Chaser dragonfly copulation wheel.

placed the giant dragonflies at a reproductive disadvantage compared with smaller species that continued to use the copulation wheel. The assumption in this theory seems to be that male dragonflies, then and now, carry passive females. On today's evidence, that would be incorrect. In fact, I think it could be argued that the female bears more than her fair share of the load.

As can be seen from Figure 7.14 and HSV.68, the geometry of the copulation wheel results in an upward tilt in the thorax of the female and a corresponding tilt in the stroke plane towards the horizontal, together with an increase in the angle of attack of the wings. The precise consequences of this are hard to determine without experimental modelling. The fact that her centre of gravity lies at least level with, and possible higher than, that of the male indicates that she is at least holding her own as far as lift is concerned. Arguably, because of the unusually steep angle of attack forced upon her wings by the wheel configuration, she could be operating her wings very close to the boundary between maximum possible lift generation and stalling.

The reproductive biology of damselflies follows rules of its own and in my opinion gives rise to some of the wonders of insect flight. My own interest in the subject was prompted by observing the strange hovering behaviour of male Common Blue Damselflies. No less than dragonflies – in fact, far more so – mating pairs are forced to run the gauntlet of pestering males. When the opportunity arises, the pair settles on surface vegetation – Canadian pondweed seems to be a favourite for this species – and the female curls her abdomen forward beneath the weed to begin oviposition. The first thing that puzzled me was the stance taken up by the male, who steadfastly remained anchored to his mate and began stationary hovering, in the process giving a remarkable impression of a miniature wind-turbine (Figures 7.17–7.18, HSV.69). The behaviour was quite contagious because, on seeing one mating pair established in the routine, other pairs would often join in, creating the equivalent of a miniature wind farm (HSV.70). I wondered what the purpose of the hovering could be. Was the male keeping the female afloat? Was he

Figure 7.17 Common Blue Damselflies *Enallagma cyathigerum*. In tandem formation with stationary hovering male ('wind turbine').

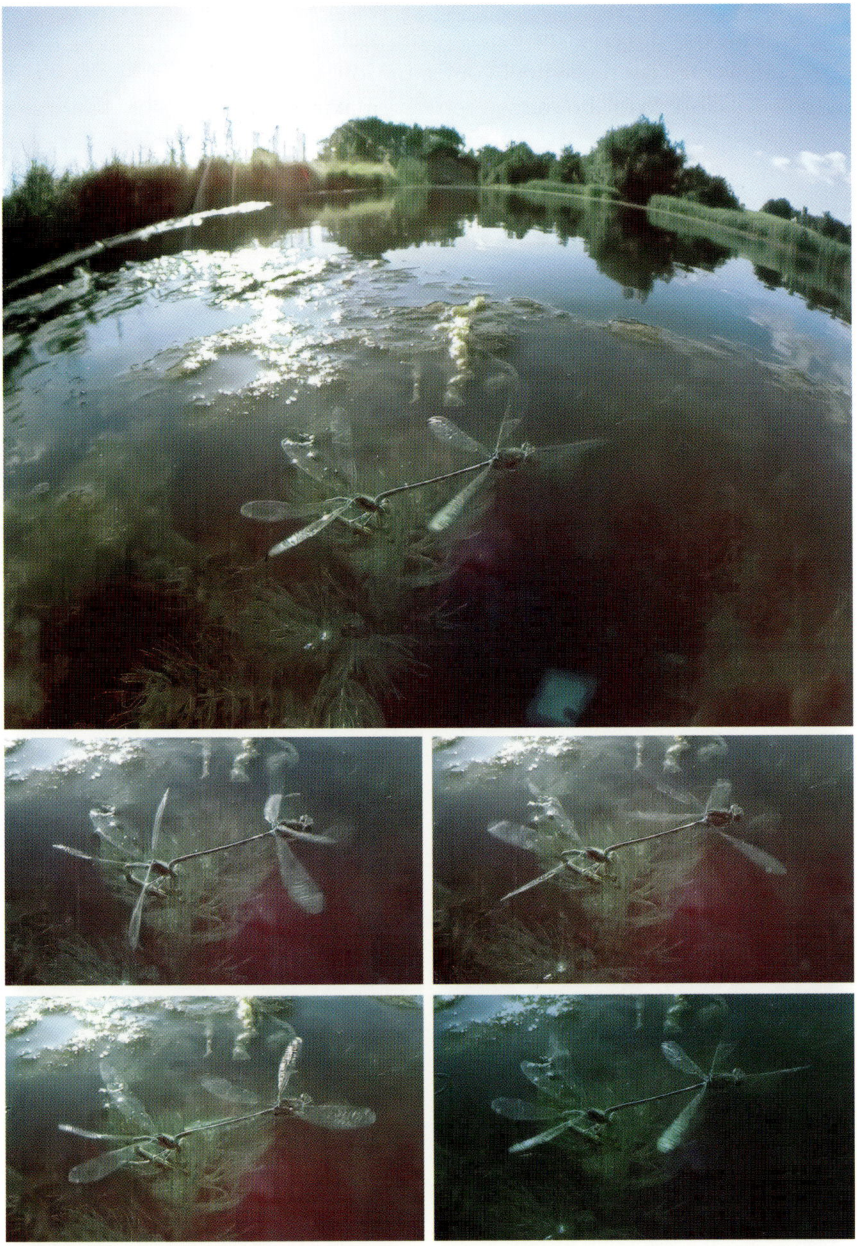

Figure 7.18 Common Blue Damselflies *Enallagma cyathigerum*. Female ovipositing while male performs stationary hovering.

saving himself from being pulled into the water? Or was he ensuring that, when the female eventually finished laying, he was ready to pull her clear of the surface and whisk her away? All these possibilities seemed far more likely than what happened.

My expectation had been that, just as in the case of any other flying insect, a female damselfly would not risk her life by immersing herself, soaking her wings, flooding her tracheal system and suffering the inevitable fate of drowning. Properly aquatic insects such as diving beetles and water boatmen beetles have evolved mechanisms to avoid this, for example by enclosing the wings inside watertight cases. I noticed that on some occasions the hovering male released the female, who then promptly disappeared beneath the surface and could be seen clambering down the waterweed. The male, now relieved of his burden, instead of leaving remained hovering over the exact spot where she had begun her descent. My immediate thought was that the female was sacrificing herself to give her eggs the best chance of survival at a location deeper in the water.

On a subsequent occasion, I scooped a submerged female towards the surface in my hand and to my surprise she instantly struggled free, broke through the surface and flew off (HSV.71). I then noticed that sometimes a tandem pair, snatching an opportunity to evade the ever-present entourage of lone males, would touch down briefly on the water surface, whereupon the female would begin her descent underwater, dragging the male with her to the point where he himself was almost submerged (Figures 7.19–7.21, HSV.72). At the last possible moment, with the water right up to his thorax, he would release her, extract himself from the surface film of the water and begin hovering over the spot, often attended by males who had witnessed the scene from nearby.

Figure 7.19 Common Blue Damselflies *Enallagma cyathigerum*. Female descending headfirst into water, pulling the attached male with her.

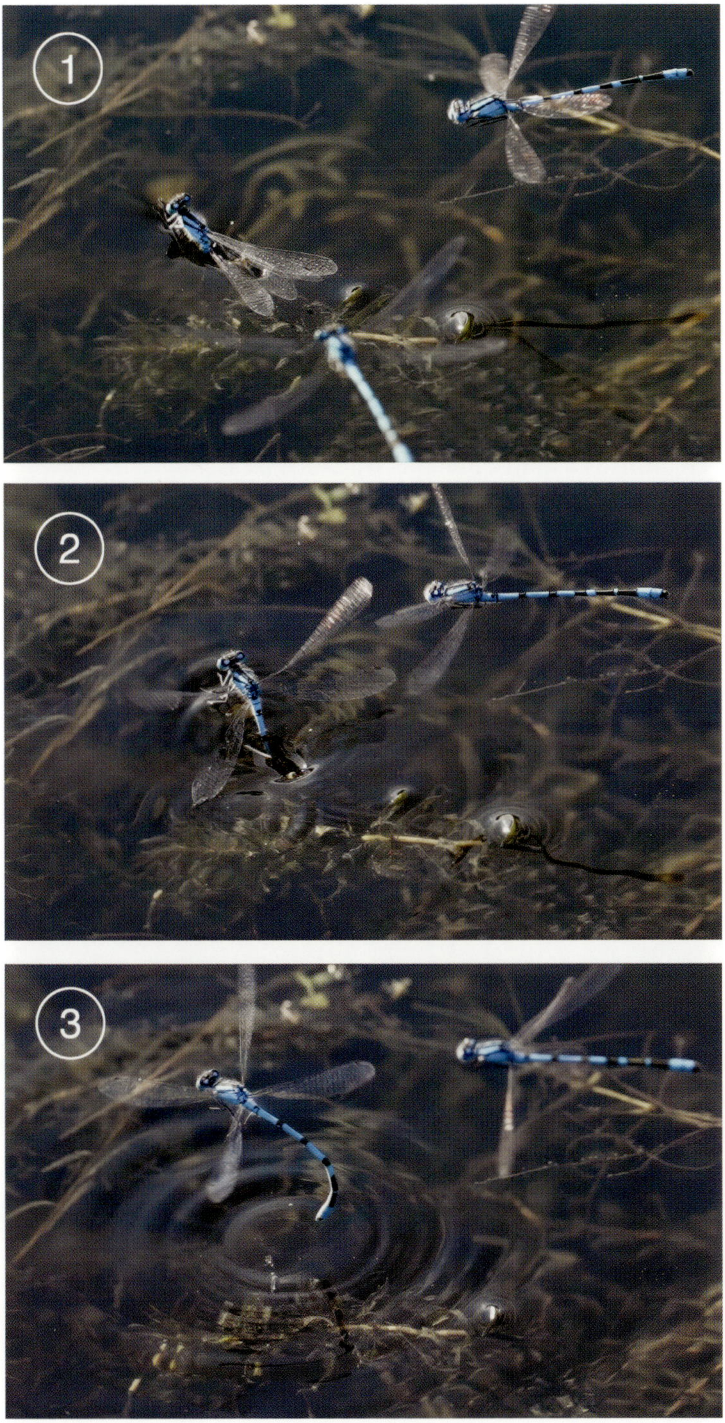

Figure 7.20 Common Blue Damselfly *Enallagma cyathigerum*. Male emerging from the water after releasing the female, who is now climbing down the waterweed.

I wondered whether the male was as adept as his partner at surviving submersion. Mating pairs can often be found resting on waterside vegetation, and it proved quite easy to grasp the wings of one of the males between finger and thumb and lower the pair into the water to a depth of about 5 cm before releasing them. I performed this experiment several times, always with the same result. The insects shot to the surface, broke free and flew off (HSV.73), all within the space of half a second.

I now understood why packs of lone males spent so much time skimming across the water: they were searching for females that had completed egg-laying and had now returned to the surface. Spotting one, a male would attempt to grasp her with his claspers and carry her off; even if he succeeded, rival males would often try to dislodge and abduct her themselves (HSV.74). If it is possible to speak of 'violence' in the context of such an apparently fragile creature as a damselfly, this was such a case.

The female survives her experience under water because her integument is water-repellent and traps a blanket of air around the body that serves as a temporary 'aqualung', functionally equivalent to the 'plastron' of aquatic beetles and bugs. The added air also increases the buoyancy of her body, which accounts for the laboured way she clambers down the waterweed: she is actively pulling her body down against the force of her own buoyancy.

Figure 7.21 Common Blue Damselfly *Enallagma cyathigerum*. Male emerging from the water after releasing the female. The scene is attended by a second, lone male and both males continue to hover over the spot where the female was deposited.

Figure 7.22a Queen Hornet *Vespa crabro* in her hibernation chamber.

Figure 7.22b Queen Common Wasp *Vespula vulgaris* in her hibernation chamber.

Buoyancy also explains the speed with which she and her partner shot to the surface in the experiment described above.

The damselfly version of courtship and egg-laying is complex, clearly successful in terms of perpetuation of the species, but inevitably full of risks for the individuals concerned. Judging by the number of bodies found floating on the surface, the female casualty rate is considerable. Seen from a strictly genetic and evolutionary perspective, however, they have fulfilled their reproductive role and are henceforth expendable.

I finish this chapter and this book with a reference to another female insect whose life does not end so abruptly following mating. In fact, it is only after a further six months or so that she finally sees the fruition of her encounter. Figures 7.22a and 7.22b show queen social wasps, a Hornet *Vespa crabro* and a Common Wasp *Vespula vulgaris* respectively, lodged in specially prepared hibernation chambers which they will occupy in a pupal-like state of dormancy throughout the winter. In the spring they will emerge, a period of renewed activity in which they will earnestly search for food to restore their depleted reserves and mature the fertilised eggs inside their bodies;

and set about establishing a new colony. It is imperative that their wings have survived perfectly intact during the period of incarceration. Notice how carefully they are folded and tucked out of harm's way beneath the body, held in place by the hind legs. This is made possible by the so-called plaiting fold which runs along the length of the wing and allows it to be folded in half when at rest.

The only other insects to possess this kind of wing fold are the solitary potter and mason wasps. On the face of it, a wing that spontaneously folds in half might seem problematic for flight. In HSV.75 a male Common Wasp flies from my hand (he has no sting and is totally harmless) and you will see how the folded wing unfolds automatically as soon as it starts beating. So, there is no conflict between flight and rest; in fact, there is evidence that the plaiting fold actively assists the process of supination, allowing the front half of the wing to hinge back on the rear half.

Worldwide there are more than a million insect species represented, in more than a thousand families. The examples of flight behaviour that I have been able to describe in this book must be only the tip of the iceberg. My experience is limited to European species, which comprise a mere fraction of the total insect fauna; but doubtless even here many more fascinating examples of flight behaviour remain to be uncovered. It is an open field for those with patience and a curious mind, no less for amateurs than for professionals.

Index

Page numbers in *italics* denote illustrations.

Adela reaumurella 144–5, *145*
aerodynamics of flight *5*, 5–6
aerofoils 5, *5*, 6
Aeshna grandis 36, 58
Aglais io 104, *105*, 106
Agrion splendens 36, *39*
aircraft 52, 57–8
albatrosses 6
Alder Fly 119, *119–21*
ambush predators 29–30
Ameles spallanziana 110, *111*
Ammophila sabulosa 54, *54*
angle of attack 5, 7, 33. *See also* stroke angle
Anthocaris cardamines *97*, 99, 103, 131, *132*
Anthophora plumipes 54, 55
antlion 6, 42, *46*
ants, worker 62
Aphelocheirus *61*, 61–2
Aphthona *123*, 123–6, *124*, *125*
aquatic bugs *61*, 61–2
Archaeopteryx 64–5
Argynnis aglaja *69*, *80*, *85*
Argynnis paphia *68*, 69, *77*, *84*
assassin bug *29*, 29–30
asynchronous wings 33–51
axillary cuticle *3*, 5

backswimmers 118, *118*
backwards flight 58–9
Banded Agrion 40, 58
Banded Demoiselle 36, *39*
banked turns 57–8, *97–8*, 99
basal sclerites *3*, *4*, 5
bees 31
Buff-tailed Bumblebee 55, *56*
Hairy-footed Flower 54, 55
Red-tailed Bumblebee *22*
beetles 111–15, *112–13*. *See also* particular species
Bibio marci *139*, 140, 148, *149*
bird-strike 101, *102*, *105*, 106
birds 6, 64–5, 111
Black-headed Cardinal Beetle *112*
Black Yellow Owlfly 42, *45*
Blue-winged Grasshopper 30–1, *51*, *108*
Bombus. *See* bumblebees
Bombus lapidarius *22*
Bombus terrestris 55, *56*
Bombylius major 54, *147*
Brimstone *32*, 33, *72–3*, *78*, *81–2*, *92–3*, *95*, *101*, *116*
Brintesia circe *84*, *86*
bristletails 108
Brown Hawker 36, 58
bumblebees *10*, *15*, 26, 55, *56*
Buff-tailed 55, *56*
Red-tailed *22*
buoyancy 162–3
Burrows, Malcolm 123
butterflies 6, 65. *See also* particular species
flight 66–106, 115–18

caddisfly *16*, *18*, 119–23, *121*, *122*
Caligo 104
camouflage 26, 30–1
Cantharis rustica *3*
capsid bug *23*
Carabidae 60
Catocala nupta *14*, *17*, 30
Cephus 129, *131*
Cercopis vulnerata 126–7
Chagas disease 30
Chalcolestes viridis 38, 58
chaser dragonflies *157*
chemical defences 28, 30
Chironomidae 140–4, *141*, *142*, *143*
chitin 1, 2
Chorthippus brunneus *109*, 110
Chrysopa perla *48*, 49
cicada *19*
Cicindela campestris *113*, 114
Cinnabar Moth *13*
circular flow 5, *5*, 36, *40*
clap and peel 9–15, 26–7, 50, 65, *86*
Clouded Yellow 131, *135*
Clytus arietis *20*
Colias croceus 131, *135*
coloration 27–31, 129, 131
Columba livia *63*, 65
Comma *69*
Common Blue, butterfly *88*, *146*, 150, *153–4*, 155

Common Blue Damselfly *34, 158–62,* 158–63
Common Darter *35*
Common Froghopper 127
Common Scorpionfly 42, *47*, 49
Common Soldier Beetle *3*
Common Wasp *21*, 28, *28*, *53*, 58–9, *163*, 163–4
Conehead Mantis *50*
Conocephalus fuscus 110
copulation 131–3, 147–64
  copulation wheel 155–8
counter-stroking 33–51
courtship behaviour 129–33. *See also* mating behaviour
cranefly *8*, 54, 148, *148*
crickets 50, 108, 110
  Great Green Bush-cricket *127*, 128

dagger fly 148–9, *151*
damage, to wings 99–106
damselflies *2*, 33–42, *36*, 155, *155*, 158–63
dancing 133, 137–46
Dark-edged Bee-fly 54, *147*
Dark Green Fritillary *69*, *80*, *85*
Darter dragonfly *42*, *43*, *57*, *156*
Diptera 31, 54, 147. *See also* particular species
display behaviour 30
Domestic Pigeon *63*, 65
dormancy *163*, 163–4
downstroke *7*, 7–8, 9–18, 65, *83–95*, 99
drag 6, 65, 138
dragonflies 33–42, *43*, *44*, 57, 58–9, 65, 155–8
  prehistoric 62, 156–7
dronefly 54, *54*, 55, *130*
dung flies 148, *150*, *151*
dynamics, laws of 5–6

egg laying 150, 155–6, 158
elasticity 1, 2, 5, 6, 8–9, 101
Ellington, Charlie 10
Empis tessellata *151*
Empusa pennata *50*
Enallagma cyathigerum *34*, *158–62*, 158–63
Eristalis tenax 54, *54*, 55, *130*
Erythromma najas 40
escape mechanisms 119–28
Eumenes coarctus *12*
European Dwarf Mantis 110, *111*
evolution of flight 60–5, 107, 109
exoskeletons *1*, 1–2
eyes 54–5, 125
eyespots 30, 104–6

Face Fly 138
Field Grasshopper *109*, 110
flash coloration 30
flea beetles *123*, 123–6, *124*, *125*
flies 31, 54, 147. *See also* particular species
flight
  butterflies 66–106
  control 52–9
  evolution of 60–5, 107, 109
  mechanics of 1–6
  take-off *96*, *97*, 101, *101*, 107–18
flight muscles *1*, 1–2
flightlessness 60–2
flowers 60
forces, on wing 6
Four-spotted Chaser *44*
free-fall 107
froghoppers 126–7

Gatekeeper 149
Geometra papilionaria *15*
geometrid moths 61
gnats 140–4, *141*, *142*, *143*
Gonepteryx rhamni *32*, 33, *72–3*, *78*, *81–2*, *92–3*, *95*, *101*, *116*
Grand Antlion 42, *46*
grasshoppers 50–1, 108–10
  Blue-winged 30–1, *51*, *108*
Great Banded Grayling *84*, *86*
Great Green Bush-cricket *127*, 128
Green Tiger Beetle *113*, 114
Green-veined White *79*, *91*, *94*, *98*, 99, 131, *133–4*, 149, *152–3*
Greenbottle 148
ground beetles 60

haemolymph 5
Hairy-footed Flower Bee 54, 55
Harlequin Ladybird *14*, *25*, *26*
Harmonia axyridis *14*, *25*, *26*
heat 27
Heath Potter Wasp *12*
helicopters 52
Hemipepsis mauritanica 28–9, *29*
hibernation *163*, 163–4
hindwings 70
Hornet *163*, 163–4
Housefly 111, *111*
hoverflies 58
hovering 36, 38, *40*, 52–9, 158–60
hunting 38, 40, *41*, 42, 56–7, 59

ichneumon wasp *9*, *11*, *12*, *19*
indirect flight mechanism *1*, 1–2
injuries, to wings 99–106

instability 52
Iphiclides podalirius *71, 86*
Iris oratoria 30, *31, 49*
Issoria lathonia *75*

jumping 107–11
escape mechanisms 119–28

lacewing *22, 48,* 49
ladybirds 26–7, 30
Harlequin *14, 25, 26*
Large Conehead *51*
Large Emerald moth *15*
Large Skipper *76*
Large White *73, 76, 83, 87, 92,* 101, *117,* 131, *132*
leafhoppers *126,* 126–7, 128
leaping. *See* jumping
Libelloides longicornis 42, *45*
Libellula quadrimaculata *44*
lift 6, 9–10, 11, 18, 33, 36
Lighthill, Sir James 10
Locusta migratoria *51*
Long-winged Conehead 110
longhorn beetle *4*
Longhorn moth 144–5, *145*
Lucilia 148

males, flight behaviour 129–46
Maniola jurtina *87, 90,* 149
manoeuvrability 57
mantids 50
Marbled White 67–8, *78, 79, 89–90,* 93, *101, 115, 117,* 133
Masked Hunter bug 30
mating behaviour 131–3, 147–64
mayfly 144, *144*
Meadow Brown *87, 90,* 149
Mediterranean Mantis 30, *31, 49*
Melanargia galathea 67–8, *78, 79, 89–90,* 93, *101, 115, 117,* 133
Mesembrina meridiana 58
Metoecus paradoxus *20*
midges 140–4, *141, 142, 143*
momentum 5–6
moths
Cinnabar *13*
geometrid 61
Large Emerald *15*
Longhorn micro-moth 144–5, *145*
plume 20, *24, 26*
Red Underwing *14, 17,* 30
Six-spot Burnet *27, 28, 28*
movement of wings 7–27, 70–99
Musca autumnalis 138

muscles, flight *1,* 1–2
Myathropa florea 58

nervous system 67
Netelia testacea *12*
Newton, Sir Isaac 5–6
Noon Fly 58
Notonecta maculata 118, *118*

Ochlodes venatus *76*
Oecanthus pellucens 110, *110*
Oedipoda caerulescens 30–1, *51, 108*
Ophion luteus *9, 11, 12, 19*
Orange-tip *97,* 99, 103, 131, *132*
owl butterfly 104, *104*
owlfly 27, 42, *45*

Painted Lady *17, 74, 80, 88, 91, 96, 103,* 134, *136*
Palpares libelluloides 42, *46*
Panorpa communis 42, *47,* 49
parallax 66–7
parallel-stroking 33, 36
Pararge aegeria 134
parasitic wasps 138, *138, 139*
patterns of wing motion 7–27, 70–99
Peacock 104, *105,* 106
Pearly Green Lacewing *48,* 49
peeling 9–15, 26–7, 50, 65, *86*
pheromones 129
Philaenus spumarius 127
Phyllotreta 125
Pieris brassicae *73, 76, 83, 87, 92,* 101, *117,* 131, *132*
Pieris napi *79, 91, 94, 98,* 99, 131, *133–4,* 149, *152–3*
Pieris rapae *77, 85, 100, 102,* 149, 150, *152*
plates, of exoskeleton *1,* 1–2
plume moths 20, *24, 26*
Polygonia c-album *69*
Polyommatus icarus *88, 146,* 150, *153–4,* 155
predation 38, 40, *41,* 42, 56–7, 59
preening 31–3
prey recognition 57
propeller effect 9, 18, 26, 33–6, 42–9, 65, 70
pterothorax 1
Pyrochroa coccinea *112*

Queen of Spain Fritillary *75*

Raphidia notata *48,* 49
Red Admiral *68, 71, 72, 81, 82, 89,* 134–7
Red-banded Sand Wasp 54, *54*
Red-eyed Damselfly 40

Red-tailed Bumblebee *22*
Red Underwing moth *14*, *17*, 30
Reduvius personatus 30
Rhynocoris iracundus *29*, 29–30
rivalry 129–33
robber fly 140, *140*
roosting 67–8
rotational flow 5, *5*, 36, *40*
Ruspolia nitidula *51*

St Mark's Fly *139*, 140, 148, *149*
sawflies 129, *131*
Scarce Swallowtail *71*, *86*
Scathophaga stercoraria 148, *150*, *151*
Sceliphron destillatorium *21*
scorpionfly 42, *47*
sexual selection 129
Sialis lutaria 119, *119–21*
Silver-washed Fritillary *68*, 69, *77*, *84*
Six-spot Burnet *27*, 28, *28*
skew thorax 38
Small White *77*, *85*, *100*, *102*, 149, 150, *152*
Snakefly *48*, 49
Speckled Wood 134
spider-hunting wasp *21*
spiders 30
spring propulsion 123–8
springtails *107*, 107–8
stability 52
startle response 30
stiffness 2
streak photography 142, *142*
stroke angle 7, 26, 70, 142
swarming 137–8
Sympetrum *42*
Sympetrum striolatum *35*

take-off *96*, *97*, 101, *101*, 107–18
Tarantula Hawk 28–9, *29*
territorial behaviour 133–6
Tettigonia viridissima *127*, 128
thermoregulation 27
thoracic segments *1*, 1–2
thorax 38
threat displays 30
tiger beetles *113*, 114–15
Tomocerus *107*, 107–8
Tree Cricket 110, *110*
Trichocera *137*, 138
turning 57–8, *97–8*, 99
Tyria jacobaeae *13*

upstroke *7*, 7–8, 18–27, *21–4*, 65, 70, *71–81*

Vanessa atalanta *68*, *71*, *72*, *81*, *82*, *89*, 134–7
veins 5
Vespa crabro *163*, 163–4
Vespula vulgaris *21*, 28, *28*, *53*, 58–9, *163*, 163–4
vision 54–5, 125–6
vortices *17*, 18

wakes 5, *17*, 18
Wallace, Alfred Russel 27
warning coloration 27–31
Wasp Beetle *20*
Wasp Nest Beetle *20*
wasps
  Common *21*, 28, *28*, *53*, 58–9, *163*, 163–4
  Heath Potter *12*
  ichneumon *9*, *11*, *12*, *19*
  parasitic 138, *138*, *139*
  Red-banded Sand 54, *54*
  Spider-hunting *21*
  Tarantula Hawk 28–9, *29*
Weis-Fogh, Torkel 9–10
whites (butterflies) 69–70
Willow Emerald Damselfly 38, 58
'windmill' configuration 33, *36*, 42, *46*, *48*, 49
wingbeat cycles
  asynchronous and counter-stroking 33–51
  patterns of motion 7–27, 65, 70–99
wingless species 60–2
wings
  attachment to thorax 2–5, *3*, *4*
  birds 64–5
  butterflies 70–106
  damage 99–106
  preening 31–3
  warning coloration 27–31
winter gnats *137*, 138
worker ants 62

Yellow Dung Fly 148, *150*, *151*

Zygaena filipendula *27*, 28, *28*